Automotive Carbon Fiber Composites

Other SAE books of interest:

Care and Repair of Advanced Composites
By William Cole, Keith B. Armstrong, & Graham Bevan
(Product Code: R-336)

Engineered Tribological Composites: The Art of Friction Material Development
By Roy Cox
(Product Code: R-401)

Dictionary of Materials and Testing
By Joan Tomsic
(Product Code: R-257)

For more information or to order a book, contact SAE International at 400 Commonwealth Drive, Warrendale, PA 15096-0001, USA; phone 877-606-7323 (U.S. and Canada only) or 724-776-4970 (outside U.S. and Canada); fax 724-776-0790; e-mail CustomerService@sae.org; website http://books.sae.org.

Automotive Carbon Fiber Composites:

From Evolution to Implementation

By Jackie D. Rehkopf

Warrendale, PA, USA

400 Commonwealth Drive
Warrendale, PA 15096-0001 USA
E-mail: CustomerService@sae.org
Phone: 877-606-7323 (inside USA and Canada)
      724-776-4970 (outside USA)
Fax: 724-776-1615

ISBN 978-0-7680-3495-0
Library of Congress Catalog Number 2011942089
SAE Order Number T-124
DOI 10.4271/T-124

To purchase bulk quantities, please contact:
SAE Customer Service
E-mail: CustomerService@sae.org
Phone: 877-606-7323 (inside USA and Canada)
      724-776-4970 (outside USA)
Fax: 724-776-1615

Visit the SAE Bookstore at http://store.sae.org

Contents

Executive Summary.. ix

Chapter 1
Introduction 1
   1.1   A Brief History of Carbon Fiber Composites................................2
   1.2   Carbon Fiber Implementation Timeline 2
   1.3   Closing Thoughts...11

Chapter 2
Carbon Fiber Composite Constituents: Fiber and Resin Types
for Automotive Applications... 17
   2.1   Carbon Fiber Primer...18
         2.1.1 Carbon Fiber History...20
         2.1.2 Carbon Fiber Developments 22
         2.1.3 Developments in Conversion Processes and
                Post-Treatments 28
         2.1.4 Future Developments Around the Carbon Fiber............29
   2.2   Resins for the Automotive Industry 29
         2.2.1 Characteristics of Different Resins Used in CFCs 30
         2.2.2 Resin Developments 31
         2.2.3 Resins with Reduced Cure Time...................................31
         2.2.4 Resins with Targeted Performance 32
         2.2.5 Resins for Targeted Processing Methods 33
   2.3   Closing Thoughts...34

Chapter 3
Carbon Fiber Composite Construction 39
   3.1   The Splendid Variety of CFCs for Automotive Applications 40
   3.2   Fiber Reinforcement Forms..42
         3.2.1 Bonded..42
         3.2.2 Unidirectional Tapes 42
         3.2.3 Stitched 42
         3.2.4 Knits 43
         3.2.5 Wovens..43
         3.2.6 Braids 45

3.3   Constructions 46

      3.3.1 Laminates 46

      3.3.2 Filament Winding 47

      3.3.3 Stitching Preforms.......................................48

      3.3.4 Three-Dimensional Braids 48

      3.3.5 Nanostitching and Fuzzy Fibers 49

3.4   Closing Thoughts..50

Chapter 4
Manufacturing Processes for Carbon Fiber Composites............53

4.1   Introduction...54

4.2   Injection Molding 55

4.3   Compression Molding 56

4.4   Thermoforming..56

4. 5   Sheet and Strand Molding Compound 57

4.6   Spray Forming...57

4.7   Pultrusion 57

4.8   Filament Winding 58

4.9   Resin Infusion Processes 59

4.10   Out-of-Autoclave Processing of Structural Components...........60

4.11   Quickstep® Process..62

4.12   Preforming Processes 62

4.13   Other Processes 63

4.14   Tooling 63

4.15   Closing Thoughts...65

Chapter 5
Machining and Joining 69

5.1   Machining 70

      5.1.1 Tool Drilling..70

      5.1.2 Tool Wear Compensation 73

      5.1.3 Tool Cutting/Trimming 74

      5.1.4 Abrasive Waterjet Cutting............................74

      5.1.5 Laser Cutting..76

5.2   Joining 76

      5.2.1 Bonding 76

      5.2.2 Mechanical Fasteners 79

5.3   Closing Thoughts...82

Chapter 6
Reclaiming/Recycling Carbon Fiber Composites 85

6.1   Introduction...86
6.2   Reclaiming and Recycling CFCs...88
6.3   Implementation of Recycled CFCs.....................................90
6.4   Closing Thoughts..92

Chapter 7
Implementation and Longevity 95

7.1   Design and Modeling 96
       7.1.1 Fiber Orientation and Composite Construction 96
       7.1.2 Computer Aided Engineering...97
       7.1.3 Mechanical Behavior of CFCs..97
7.2   Physical Testing 99
7.3   Quality Control...100
7.4   Non-Destructive Evaluation...101
7.5   Repair 104
7.6   Closing Thoughts..105

Chapter 8
Concluding Thoughts 107

8.1   Manufacturing and Assembly in Legacy Plants 108
8.2   Advancing with the Advancements of Other Materials 109
8.3   Industry and Public Acceptance.....................................113
8.4   Closing Thoughts..115

About the Author...117

Executive Summary

The advantages of carbon fiber composites (CFCs) in automotive design are high stiffness, high specific strength (strength-to-weight ratio), excellent fatigue endurance, corrosion resistance, generally good impact resistance, and flexibility in design that permits them to be tailored to design requirements. Composites also facilitate a lower parts count by reducing the number of subassemblies and fasteners. Replacing metal with CFCs can provide significant weight reduction, which has become particularly important in our current society that is facing high fuel prices and much more stringent emissions standards. Aside from passenger vehicles, heavy-duty on-highway trucks and military vehicles are also exploring the use of CFC components to enable better fuel economy and/or increase payload. Unfortunately, CFCs also have notable disadvantages, including relatively high material and fabrication costs, poor compressive and shear properties, and the necessity for non-destructive inspection techniques to detect flaws or damage.

The main factors in the automotive industry driving fiber development and resin development center around cost, performance, cure time, and processing method. The years 2010 and 2011 have seen an incredible amount of cooperation and partnerships between companies operating at different points in the value stream to bring new materials and processing technologies to market quicker. Carbon fibers and matrix resins, and their developments, are not independent of the other aspects of manufacturing a carbon fiber composite component, nor are they necessarily independent of each other. Different resins process differently with regard to the time, temperature, and pressure required for fiber wet-out and consolidation. Additionally, different fiber-resin constructions require different processing methods. Selecting the right fiber, resin, and construction for a particular application requires knowledge of not only the fiber and resin material properties, but also of the method of manufacturing. The method of manufacturing influences the composite construction and end properties, while the surface quality of the finished part (Class A or non-Class A) and the production volumes to be made in turn dictate what manufacturing methods are technically and economically viable. One must also factor in the commercial competitiveness with other materials with regard to vehicle installation, maintenance, and lifecycle issues. There is usually more than one way to make a carbon fiber composite automotive part, and all factors should be considered to make the best decision.

Both the aerospace and automotive industries are driving changes in carbon fiber composite technology to produce components that have lower material cost and targeted performance. Those developments will in turn lead to some developments in the manufacturing processes. For example, changes in processing temperatures can result in changes in tooling materials and heat sources, and changes in composite construction can lead to changes in material handling during component manufacture. Such future advancements driven by raw material and construction improvements will be additive to the advancements driven directly by the manufacturing process to improve areas such as part-to-part cycle time and energy efficiency.

One of the most challenging aspects of implementing CFC components in vehicle design is attaching them to the rest of the vehicle. This usually requires machining and joining, which must be done in a manner that retains the mechanical properties of the CFC component as well as possible, provides a strong and durable joint, is cost-effective, and fits with the OEM assembly process and vehicle production rate.

Worldwide, automotive companies are facing some challenging energy and environmental issues. In the U.S., the 2010 Corporate Average Fuel Economy (CAFE) regulation that increased fuel economy from 27 to 35 miles per gallon by 2016 has already resulted in concerted efforts to implement more lightweighting materials, including CFCs, in vehicle design. Means to recycle CFCs and other lightweighting materials must be developed to just maintain the current level of recyclability of vehicles made predominantly with steel. The CFC recycling industry is still in its infancy and the processes are expensive and complicated. The industry has formidable requirements, including consistent scrap availability, appropriate size reduction technologies, established process parameters, the infrastructure for material collection, and standardization of recyclate properties. The technical and economic issues with recycling/reusing CFCs are best developed during vehicle design to aid in both the recovery of the material as well as potential implementation of the recyclate back into a vehicle.

Implementation and longevity of CFC components in mainstream vehicles hinge on a multitude of technical issues, covering raw materials, fabrication, assembly to other (CFC or non-CFC) components, and in-vehicle performance. However, addressing all the technical issues will not guarantee first-use or long lasting use of CFCs in mainstream vehicles. Acceptance of the material is also key — the acceptance of CFCs by OEMs through the inclusion of CFCs in their portfolio of materials from which they can design mainstream vehicles, and acceptance by the consumers with regard to cost and performance throughout the vehicle life, which inevitably includes damage and repair.

This is an exciting time for the carbon fiber composites and automotive industries. The current need to drastically lightweight the U.S. vehicle fleet in the next few years provides a great opportunity for CFCs to find prominence in mainstream vehicles. Their advantageous high specific modulus and strength can result in weight savings up to 60% compared to conventional steel designs. However, before CFCs find prominence, significant inroads in reducing the relatively high material and fabrication costs, long part-to-part cycle times, and slow assembly/attachment to other vehicle components will need to be made. Progress will also be needed in the areas of damage detection, repairability/replaceability, and recycling.

The combined chapters of this book highlight current activities surrounding automotive carbon fiber composites and the anticipated direction of developments in the next 5-10 years. The objective is to provide a high-level view as opposed to technical treatises, preparing the reader for meaningful discussions with composites engineers and technicians, fiber suppliers, resin suppliers, tool and equipment manufacturers, as well as business development and lifecycle workers. The possibilities of carbon fiber composites in automotive applications are plentiful — and more promising than ever before in the history of the automobile.

Chapter One

Introduction

We must dare to think "unthinkable" thoughts.
We must learn to explore all the options and
possibilities that confront us in a complex and
rapidly changing world.
—James W. Fulbright

1.1 A Brief History of Carbon Fiber Composites

In the early 1940s, the defense industry spawned the industrialization of fiber reinforced plastics (FRPs), particularly for use in aerospace and naval applications. The U.S. Air Force and Navy capitalized on FRP composites' high strength-to-weight and inherent resistance to weather and the corrosive effects of salt air and sea. The rapid development and use of composite materials beginning in the 1940s had three main driving forces: 1) Military vehicles, such as airplanes, helicopters, and rockets, placed a premium on high-strength, lightweight materials; 2) The emergence of new, lightweight polymers offered solutions for a variety of uses, provided that something could be done to increase the mechanical properties of plastics; and 3) The extremely high theoretical strength of certain materials, such as glass fibers and carbon fibers, was being discovered to solve the problems posed by the military's demands.

The 1950s brought new revolutionary applications for FRP composites. The same technology that produced the reinforced plastic for the Manhattan Project in World War II sparked the development of high-performance carbon fiber composite materials for solid rocket motor cases and tanks. Carbon fibers have low heat expansion, high dimensional stability, high tensile modulus and strength, and they sustain these excellent mechanical properties under high temperatures.

The high potential strength of carbon fiber was realized in 1963 through a process developed at the Royal Aircraft Establishment at Farnborough, Hampshire. The process was patented by the UK Ministry of Defence, then licensed by the National Research Development Corporation (NRDC) to three British companies: Rolls-Royce, which was already making carbon fiber, Morganite, and Courtaulds. They established industrial carbon fiber production facilities within a few years. Rolls-Royce took advantage of the carbon fiber composite properties to break into the American market with its RB-211 aero-engine using carbon fiber in the engine's compressor blades. Unfortunately, during testing they proved vulnerable to damage from bird impact. In 1968 Rolls-Royce's ambitious schedule for the RB-211 was endangered, and the problems became so great that the British government nationalized the company in 1971 and sold off the carbon fiber production plant to form Bristol Composites.

Although the RB-211 carbon fiber blades were not a success in the early 1970s, the carbon fiber technology was born and its long journey to maturity was under way.

1.2 Carbon Fiber Implementation Timeline

Aerospace

When it first flew in 1968, the all-metal Boeing 747 weighed 750,000 pounds. Being so massive, it required a tremendous amount of fuel to lift and propel it. Any weight reduction that could be achieved through new materials was welcomed, and engineers at the Langley Laboratory believed they could provide that with their major advancement in carbon fiber composites (CFCs). Efforts to incorporate these new materials had been underway for several years, but the cost of research and development to be able to flight-test them and obtain durability data, as well as the anticipated fabrication costs for production were major hurdles. Composite development became the focus of the 1972 joint Air Force-NASA program called Long Range Planning Study for Composites (RECAST), and in 1976 the Composite Primary Aircraft Structures (CPAS) program became one of six main programs of the Aircraft Energy Efficiency (ACEE) project that was generated by the energy crisis of the 1970s. The CPAS program included industry partners Boeing Commercial Airplane, Douglas Aircraft, and Lockheed, and was coordinated by Langley Research Center.

The objective of CPAS was to provide the technology and confidence for commercial transport manufacturers to commit to production of composites in future aircraft, targeting to achieve a 25% weight reduction and 10-15% fuel savings. The technology included design concepts and cost-efficient manufacturing processes. The confidence would come with proof of the composite's durability, cost verification, U.S. Federal Aviation Administration (FAA) certification, and ultimately its acceptance by the airlines. Using composites for the wings and fuselage promised the greatest savings, but also were the most technically challenging because these components were so vital to aircraft safety. To overcome some of the uncertainties of the materials, secondary structures (upper aft rudders, inboard aileron, and elevators) were the first candidates for composite materials. Once these were successful, then the development of medium primary structures (vertical stabilizer, vertical fin, and horizontal stabilizer) would begin.

The upper aft rudder on the Douglas DC-10 was one of the first secondary structures studied (work had been initiated prior to the ACEE but was completed within the ACEE). The next secondary structures designed were elevators; 10 units were made for the Boeing 727 and flight-testing began in March 1980. Boeing approved the elevators, made with Toray T300, for use on the 757 and 767. The final secondary structures developed were the inboard ailerons; eight units installed on the Lockheed L-1011 airplane began flight testing in 1982. The design of the vertical fin (medium primary structure) for the Lockheed L-1011 was transferred to the ACEE program in 1976 after having been initiated in 1975. The composite materials failed prior to reaching ultimate load, and development ceased under the ACEE. The horizontal stabilizer designed for the Boeing 737 also experienced structural failures during ground tests, but they were corrected and received FAA certification in August 1982 and hit first commercial flight in April 1984. The vertical stabilizer for the Douglas DC-10, redesigned after initial ground failure, was FAA certified in 1986 and hit first commercial flight in January 1987.

The ACEE composites program lasted 10 years, from 1976 to 1985. It ended before achieving its major goal of developing wings and fuselages with composite materials, the stated goal of the program because the wing and fuselage represented 75% of the weight of the airplane. Wings and fuselages made of composites would have achieved significant weight savings and fuel economy. There were several reasons that these were never developed by ACEE, including the amount of time and resources devoted to assessing the risk of carbon fiber. Unanticipated at the start of the project, this potential risk became a serious threat to the use of composites and it became necessary to prove that there was little risk in their use. After doing so, in 1981 NASA was finally able to devote all of its attention to wings and fuselages, but by then engineers were approaching their development differently than they had the previous components. Whereas before they had developed composites that replaced entire metal components on aircraft, NASA now decided to try to incorporate composite pieces into the fuselage (a section barrel) and wing (short-span wing box). Boeing studied the damage tolerance of composite wings, the threat posed by lightning strikes, and their fuel sealing capabilities. Lockheed examined acoustic issues, such as how noise was transmitted through flat, angular, composite panels and how to reduce it. By this time, the ACEE program and its funding were nearly at an end, so the ultimate goal of composite wings and fuselages was never attained. Nevertheless, the success ACEE had with secondary components was almost revolutionary. This ten-year period had been the golden age of composites research in the United States, with the ACEE's achievements in secondary structures being vital in implementing a new type of material as an alternative to the traditional metals used in airplanes.

The Composite Primary Aircraft Structures program had several very significant results over its lifespan. It produced 600 technical reports and provided a cost estimate for developing these materials and a confidence in their durability and long-term use. Composites received certification by the FAA, as well as general acceptance by the airline industry. Overall, it is estimated that the ACEE program accelerated the use of composites in the airline industry by at least five to ten years. Prior to the ACEE program, aircraft manufacturers were reluctant to investigate the opportunities these composites offered because of costs and unknown performance capabilities. The NASA Aircraft Efficiency Program provided the experience and confidence needed for extensive use of composites in future aircraft. By the 1990s, these composite materials resulted in a fuel efficiency savings of 15%. Since the ACEE program ended, CFC use has increased, though not as dramatically as anticipated. While the weight savings and fuel efficiency were attractive, their implementation was hindered by the high cost of producing and certifying them for flight compared with metal. The interested reader is referred to Bowles [2010] for the well-written and thorough review of the development of composites during NASA's Aircraft Energy Efficiency Program.

The new Boeing 787 Dreamliner, provided to its first customer in September 2011, is the first major commercial aircraft with composites comprising the majority of its materials, as 50% of primary structures, including fuselage and wing, were to be composites. The Airbus A350 XWB also contains more than 50% composites by structural weight [Keynote 2010]. The ACEE program can be credited for taking materials that were exotic and untested, and transforming them into usable structures on commercial and military aircraft.

Military

While the commercial aircraft were making strides in implementing CFCs, so was the military, and somewhat more aggressively. CFC activity in military applications began in the late 1960s in the UK and the USA, and although there have been slower periods over the decades, it has essentially been progressing ever since. Composites first appeared in fighter and attack planes such as the F/A-18 C/D and AV-8B II, which require high thrust-to-weight ratios, and more recently in stealth planes such as the B2 Spirit. Composites typically comprise 20–50% of the aircraft structure for modern fighters such as the F-22, F-35 Joint Strike Fighter, Rafale, and Eurofighter. Many of these aircraft also have engines that incorporate CFCs [AeroStrategy 2006]. CFCs are also used in rotary aircraft, in the rotor blades as well as the structure. Further, the Airbus A400M military transport introduced in 2010 has an all-composite wing and total composite content of 35% of the aircraft's empty weight.

CFCs have cascaded to military ground and marine vehicles, as they are driven by the same need for lower weight and low radar signature. CFCs also see some application, often in conjunction with other materials, to provide ballistic and blast protection.

Racing Cars, Supercars, and Niche Vehicles

Racing cars used to be made from the same material as road cars, namely steel, aluminum, and other metals. It was inevitable that the automotive industry would take notice and begin adopting composite materials, and in the early 1980s Formula 1 cars underwent an important change: the use of carbon fiber composite materials to build the chassis.

Back in the 1960s, Colin Chapman, chief designer of Lotus, introduced the monocoque to Formula 1 by placing thin plates around the bars of the chassis. This new semi-monocoque design increased the stiffness of the chassis. In the 1970s, a typical monocoque was made from aluminum and contained up to 50 different sections. When these structures proved to have inadequate strength to manage down-forces from the wings, John Barnard from McLaren developed the first self-supporting chassis made from carbon fiber. The MP4-1 (initially known as the MP4) chassis was constructed from just five carbon fiber moldings. The American high-performance composites provider Hercules Aerospace produced the chassis for McLaren, which did not have the materials knowledge at the time. The MP4-1 debuted in Formula 1 in 1981, and it quickly revolutionized car design in this racing series with new levels of rigidity and driver protection. Within months the design had been copied by many of McLaren's rivals. The MP4-1 brought McLaren six wins: one in 1981 (MP4), four in 1982 (MP4B), and a further victory in 1983 (MP4/1C) [Company: McLaren 2011]. In 1981, the McLaren drivers proved the safety and advantages of this new carbon fiber composite construction. John Watson walked away from a crash at the 1981 Italian Grand Prix that nearly destroyed the MP4, and teammate Andrea De Cesaris survived over 20 accidents in the 1981 season alone. Today, CFC is the standard construction material for every car in the Formula 1 grid [Chapman 2009].

Today, most racing car chassis, which includes the monocoque, suspension, wings, and engine cover, are built with carbon fiber. The material has been cascaded to supercars and other niche vehicles. The super-car Ascari Kz1, which was introduced in 2003, incorporates a carbon fiber chassis and body.

The first road car to use a carbon fiber structure was the three-seater McLaren F1 in 1992. This million dollar F1 road car chassis took 3000 hours to make. In 1997, the Mercedes-Benz CLK-GTR super-sports car was presented as a road-registered vehicle with a carbon fiber body [Mercedes-Benz 2011]. In 2004, the Mercedes SLR McLaren began a six-year production run of some 2150 vehicles. Each road car chassis took 400 hours to make, a dramatic reduction over the F1 road car chassis.

The 2012 McLaren MP4-12C (shown in Fig. 1.1), with a price of $231,400, has a projected production run of 1000 vehicles in 2011 and more than 2000 vehicles in 2012. The 12C chassis, which the company calls a MonoCell, is produced from non-crimp carbon fiber fabric using resin transfer molding. The MonoCell, or "tub" that encloses the driver, weighs 165 lb (75 kg) and is produced in four hours, or 1/100th of the time it took to make the Mercedes SLR McLaren chassis in the 1990s. McLaren's group manager Claudio Santoni ponders whether this is the start of a movement

Figure 1.1. Claudio Santoni, Function Group Manager, Body Structures, McLaren, sits in a rolling chassis of the 2012 McLaren MP4-12C. The two-seater's 592-horsepower twin-turbo V-8 will launch it to terminal velocity of 205 mph — at about half the cost of any previous effort [Mayersohn 2011].
New York Times News Service

toward carbon fiber applications in mainstream vehicles, stating "I keep thinking in 30 years that the VW Golf would be possible." [Mayersohn 2011]

In 1992, Ferrari Enzo joined the extreme performance car group, with an advanced composite body work (some CF) and a CF-aluminum honeycomb sandwich chassis. Only 400 Enzos were produced [Ferrari 2011].

The Porsche Carrera GT sports car, which began sales in December 2003, won the car category of Popular Science Magazine's 2003 16th Annual "Best of What's New" for its advanced technology and development of its chassis, which is comprised mainly of CFC with a honeycomb intermediate layer of aluminum. The chassis contained more than a thousand CFC sections, up to 10 layers thick in spots and of different fabric forms [Porsche 2003].

Lamborghini has also employed CFCs for a number of years. The use of CFCs in the Lamborghini Murcielago has been discussed in papers by Ferraboli. The Murcielago is a two-door, two-seat sports car produced between 2002 and 2010. It has been documented [Masini and Feraboli 2003; Masini, Bonfatti, and Feraboli 2006] that the car features a CFC body except for the roof structure (i.e., bumpers, fenders, and hood) and structural components such as the transmission tunnel, floor pans, and rocker panels. The Roadster has an entire carbon/epoxy body excluding the doors, roof, and pillars, and has carbon/epoxy chassis components including the transmission tunnel, floor pan, wheel housings, doorsill stiffeners, and engine bay subframe.

The Lamborghini Aventador has a CFC monocoque, while the Lamborghini Gallardo LP 570-4 Superleggera has some CFC body elements, CFC engine cover, and CFC rear spoiler [Lamborghini 2011]. The Lamborghini Sesto Elemento concept vehicle features forged CFC parts for its monocoque body, front- and rear-end structures, suspension components, interior parts, and exterior body panels. This new forging process was developed at the University of Washington in Seattle, and involves high-pressure molding of a system of resin and discontinuous carbon fibers, one or two inches long. It takes about three to five minutes to make a part. Compared to traditionally made CFC parts, the forged parts are not as strong but are expected to be well suited for under-the-skin parts such as the suspension and monocoque, and are expected to be orders of magnitude cheaper [Sedgwick 2011]. At Lamborghini's Advanced Composite Research Center in Sant'Agata Italy, forging of chopped fibers about 8-20 in (20-50 cm) long in a heated 1000-t press — which is one part of the extensive range of equipment — is underway for possible future automotive applications. This forged material contains 500,000 braided fiber/in^2 and is stronger than titanium but 1/3 the density [Birch 2011].

Lamborghini and Audi AG are both part of the Volkswagen Group, and the two companies enjoy substantial technical collaboration on CFCs and aluminum, from research and development on through to production [Birch 2011]. Lamborghini's 30-some years of experience with CFCs are sure to benefit the entire Volkswagen Group as automakers strive to implement the lightweighting material.

The Lotus Evora Carbon Concept, unveiled at the 80th Geneva Motorshow, features a structural CFC roof, and Lotus Motorsport-influenced CFC diffuser and splitter [Lotus 2010].

The Lotus Elan high-performance sports car, with a selling price of $118,500, is to be produced in 2013 with a carbon fiber body [Lotus Elan 2011].

Bentley's T35-4 demonstrator vehicle of low-cost technology featured direct carbon fiber preforming (DCFP) of the front end, which is then wrapped in a thin layer of metal. The process is faster than autoclave processing and is less expensive because it uses the cheapest carbon fiber available. Their DCFP has been used to produce the CFC wheel well on its all-new Mulsanne flagship sedan, and a future floor pan for an upcoming Audi vehicle. The T35-4 demonstrator vehicle also used an undisclosed composite technology from its parent Volkswagen AG for the rear-end [Amen 2010].

The Lexus LFA supersports car began production in 2010 and is slated for small production volume (< 1000 vehicles). The vehicle took 10 years to develop [Lexus. Milestones, Sugo 2011]; in May 2003 the bare chassis, fabricated from CFC, was completed [Lexus. Milestones, Bare 2011]. The vehicle has a CFC cabin and employs three CFC molding processes and an advanced new process for joining CFC and metal. The CFC construction saved 100 kg (220 lb) over an equivalent aluminum design. The CFC technology is primed for future mass production of other Lexus models [Lexus. LFA 2011].

The CFC technology was developed by the LFA Plant in Japan and will be used for future Lexus and Toyota models as well. CFC comprises 65% of the body and enabled a reduction in vehicle parts from approximately 15,000 to 12,000, with each part being assembled by hand [Lexus. Milestones, LFA 2011].

BMW i3 and i8 models scheduled for 2013 production will have a CFC passenger cell [BMW M3 2011].

The 2003 Dodge Viper used CFC for the front fender supports [Nelson 2005] and CFC sheet molding compound (SMC) augments to fiberglass SMC in the doors and windshield surrounds. The CFC SMC was selected for being more practical for the 3000-unit Viper volume than aerospace prepreg technology that was used at the time in some supercars, in volumes of only a few hundred per year. Although the Viper was a low-volume niche vehicle, it was the first U.S. production vehicle to use 100% CFC parts [Brosius 2003].

Ford's 2004 GT used CFC for the rear decklid inner structures in its 4500 vehicle run [Nelson 2005]. The CFC decklid inner and superplastically formed outer were adhesively bonded and roll-hemmed together [Koganti and Bennett 2005].

The 2004 GM LeMans Commemorative Edition Corvette Z06 used autoclaved carbon/epoxy prepreg for its 2100 exterior hood panels [Nelson 2005] . The CFC outer panel hoods were manufactured using Toray's unidirectional prepreg P3831C-190, which is comprised of their T60024K carbon fibers in their G83 quick-cure epoxy resin. (Hood inner panels were made of glass fiber SMC.) Part-to-part cycle time was three hours, including application of mold release, layup, vacuum bagging, loading the autoclave, pressurization, temperature ramp up and cure, cool down, unloading the autoclave, and part removal. This was significantly faster than the one or two parts per day that was common for autoclaved parts made for supercars. This was also the first application of CFC on a painted, Class-A exterior body panel offered as original equipment on a North American vehicle [Brosius 2004].

The 2006 Corvette Z06 continued GM's carbon fiber efforts for a larger production volume — slated to be 7000 for 2006, which was nearly seven times the volume of the 2004 LeMans edition. The same prepreg that was used in the 2004 LeMans edition was used for the 2006 CFC fenders, hood, and roof. Vermont Composites (which has since changed ownership and is now known as Plasan Carbon Composites) was able to reduce the autoclave cycle to 90 minutes through a combination of tool material changes, tool inserts, and 3-D airflow and thermal modeling [Nelson 2005].

High-performance race cars have always been viewed as a proving ground for technology transfer to passenger vehicles, and today that is more evident than ever with regard to CFCs. An apt example is Lexus, having developed the LFA over 10 years, including its own CFC technology, with the vision that the CFC technology is primed for future mass production of other Lexus models [Lexus. LFA 2011]. Another intriguing example of high-performance car technology transfer that is underway is McLaren working with Carbo Tech, an Austrian CFC firm, with plans to build 5000 cars a year at a new factory near its base in Woking [VW 2011].

Automotive "Mass" Production Vehicles

It has been 30 years since McLaren debuted its Formula 1 CFC chassis in 1981. Having a density about ¼ that of steel and ½ that of aluminum, and yield strength up to 10 times that of steel and more than 10 times that of aluminum and magnesium [Lutsey 2010], CFCs have always been attractive to the mainstream automotive industry for their potential to significantly reduce vehicle weight while maintaining or improving structural performance. A CFC part is typically 30–50% the weight of a steel part [Adam 1997] and 50% the weight of an aluminum part with equivalent strength. Mainstream automotive began toying with the material decades ago: for example, the 1979 CFC Ford LTD prototype [Brylawski and Lovins 1998] and the GM Ultralite concept vehicle in 1992 with CFC body [Gromer 1992], and the GM Precept prototype with an aluminum-carbon fiber body, produced in 2000 through the Partnership for a New Generation Vehicle (PNGV) [Robinson 2001].

There are a number of reasons why CFCs have not penetrated mainstream automotive in the past 30 years [Brylawski and Lovins 1998]. Two of the major reasons have been the traditionally high cost of CFC materials and the high labor intensity of fabricating parts with it. With a current national agenda to reduce the United States' dependency on oil, there are significant efforts underway in industry and government to significantly reduce the weight of the vehicle fleet, by up to 50%. This has led to multiple activities to drastically reduce the time and costs of fabricating and implementing CFC parts into mainstream automotive components.

CFCs really began entering low-volume mainstream vehicles in the late 1990s and early 2000s. CFC driveshafts have been used for years in the Mitsubishi Montero, Nissan Z Coupe, and Mazda RX-8. Implementation into low-volume mainstream vehicles continued throughout the 2000s. The BMW M6 Coupe has had a CFC roof panel from 2006 through 2010 [Seabrook 2005], building off its experience with a limited M3 CSL series that debuted a carbon fiber roof. The 2010 Audi R8 Spyder 5.2 had an optional carbon fiber exterior package in which the material was integrated on the lower front bumper and rear diffusers [Audi 2010] which is also now available on the 2011 Audi R8 Coupe.

The years 2010 and 2011 have seen unprecedented movement worldwide by the automotive original equipment manufacturers (OEMs) into CFC development. In April 2010, BMW announced a joint venture with SGL Group and Mitsubishi Rayon to produce polyacrylonitrile- (PAN-) based carbon fibers at a facility in Moses Lake, WA for exclusive use in developing the CFC passenger cell for its Megacity Vehicle (MCV) due on the market in 2013 [BMW's MCV 2011]. They have been working on an injection molding process to produce CFC parts in minutes. They report that the vehicles perform well in crash tests, and in some cases damaged CFC parts are repairable. Also in April 2010, Daimler AG and Toray Industries announced they will jointly develop CFC materials, starting with the Mercedes SL-Class in 2012, in their effort to implement CFC in mass-produced auto bodies. The Lexus LFA with its CFC cabin began production in December 2010, with the expectation that the technology will be cascaded to mass-production of other Lexus and Toyota models. In February 2011, Audi announced a partnership with Voith to develop CFC components that will allow Audi to mass produce lightweight vehicles [Voith 2011]. At the Geneva car show in March 2011, VW announced it will invest $194M (140M Euro) for an 8% stake in SGL. Plasan Carbon Composites, headquartered in Bennington VT, is commercializing its new high volume technology over the next couple years and is establishing a new manufacturing facility in west Michigan in early 2012. It is working with the OEMs to implement Class-A CFC body panels and validate semi-structural and structural components for base model vehicles.

In February 2010, Dow Chemical Co. announced it received a $5M grant from the Centers of Energy Excellence (COEE) program through the Michigan Economic Development Corp. (MEDC) to help accelerate the manufacture of lower-cost carbon fiber for industrial use, including automotive [Michigan 2010].

Also in February 2010, it was announced that the Australian government would invest $37M (AU) into the $120M Australian Future Fibres Research and Innovation Centre (AFFRIC) at Deakin University to position Australia as an international hub for new materials industries. Construction of the center, expected to be completed by September 2012, anticipates establishing the world's first independent carbon fiber research and development pilot plant [Sudharsan 2011; VCAMM 2010].

The importance of carbon fiber composites to the U.S. national interests is evident in the Carbon Fiber Composites Consortium established in July 2011 by Oak Ridge National Laboratory. This consortium will accelerate the development of low-cost carbon fiber and CFCs and their implementation into commercial applications in various industries, including automotive. The Consortium had 15 founding members and a total of 26 members by the first Consortium meeting; they include carbon fiber manufacturers, resin suppliers, carbon fiber composite providers, private and federal laboratories, automotive part suppliers, and a designer/manufacturer of open wheel race cars.

1.3 Closing Thoughts

The advantages of CFCs in automotive design are high stiffness, high specific strength (strength-to-weight ratio), excellent fatigue endurance, good corrosion resistance, generally good impact resistance, and flexibility in design that permits them to be tailored to design requirements. Composites also facilitate a lower parts count by reducing the number of subassemblies and fasteners. Replacing metal with CFCs can provide significant weight reduction, which has become particularly important in our current society that is facing high fuel prices and much more stringent emissions standards. Aside from passenger vehicles, heavy-duty on-highway trucks and military vehicles are also exploring the use of CFC components to enable better fuel economy and/or increase payload. For example, Meritor, the automotive supplier to commercial trucks, (formerly Arvin Meritor) is investigating the advanced material for the bowl on a rear axle, although the higher cost is an issue [Gehm 2011].

Unfortunately, CFCs also have notable disadvantages, including relatively high material and fabrication costs, poor compressive and shear properties, and the necessity for nondestructive inspection techniques to detect flaws or damage.

This book is intended to provide high-level edification on carbon fiber composites specific to the automotive industry in today's market and the market envisioned for the next 5-10 years. It is targeted to those making the decisions to consider, and implement, carbon fiber composites into automotive design, and thus highlights current activities as opposed to providing technical treatises. Upon reviewing the remaining chapters, the reader will hopefully feel armed to engage in meaningful discussions with composites engineers and technicians, fiber suppliers, resin suppliers, tool and equipment manufacturers, as well as business development and lifecycle workers, to fully consider the possibilities of carbon fiber composites in automotive applications.

References

Adam, H. 1997. "Carbon fibre in automotive applications." *Materials & Design* no. 18 (4-6, December 1, 1997):349-355.

AeroStrategy. 2006. "Material Wealth: The Growing Use of Composites is Ending the 80-Year Reign of Metallic Aircraft." *An Aviation Industry Commentary* no. September 2006 (http://www.aerostrategy.com/downloads/commentaries/commentary_oct06.pdf).

Amen, James M. 2010. "Bentley Making Cheaper, Stronger Carbon-Fibre Parts." *WardsAuto.com* no. September 16, 2010 (http://wardsauto.com/ar/bentley_carbon_fiber_100916/ Accessed August 8, 2011).

Audi. "2010 Audi R8 Spyder 5.2 Carbon Fiber Sigma Exterior Package." (http://www.audiusa.com/us/brand/en/models/2010/r8_spyder_52/features_and_specifications/equipment_packages/carbon_fiber.html).

Birch, Stuart. 2011. "Multimaterial collaboration." *Automotive Engineering International Online* no. July 5, 2011:43-46. SAE International, Warrendale, PA.

"BMW M3 CRT debuts carbon fibre technology." 2011. *Reinforced Plastics.com* no. July 6, 2011 (http://www.reinforcedplastics.com/view/19158/bmw-m3-crt-debuts-carbon-fibre-technology/).

"BMW's MCV, the first mass-produced car with a carbon passenger cell." 2011. *JEC Composites* no. February 23, 2011 (http://www.jeccomposites.com/news/composites-news/bmw-s-mcv-first-mass-produced-car-carbon-passenger-cell).

Bowles, Mark D. 2010. "The "Apollo" of Aeronautics. NASA's Aircraft Energy Efficiency Program 1973-1987." (SP-2009-574).

Brosius, Dale. 2003. "Carbon Fibre Goes Mainstream Automotive." *Composites Technology* no. April 2003 (http://www.compositesworld.com/articles/carbon-fiber-goes-mainstream-automotive Accessed August 9, 2011).

———. 2004. "Corvette gets leaner with carbon fibre hood." *High Performance Composites* no. March 2004 (http://www.compositesworld.com/articles/corvette-gets-leaner-with-carbon-fiber-hood Accessed August 9, 2011).

Brylawski, Michael M. and Amory B. Lovins. 1998. "Advanced Composites: The Car Is At The Crossroads." *Rocky Mountain Institute* (http://old.rmi.org/images/PDFs/Transportation/T98-01_CarAtCrossroads.pdf).

Chapman, Giles. 2009. "Illustrated Encyclopedia of Extraordinary Automobiles." *DK Publishing, New York, NY.*

"Company: McLaren: History & Heritage. McLaren - Motor Racing Heritage." 2011. (http://media.mclarenautomotive.com/page/9/ Accessed 2011).

Ferrari. 2011. "All Models: Enzo Ferrari." (http://www.ferrari.com/English/GT_Sport%20Cars/Classiche/All_Models/Pages/Enzo_Ferrari.aspx Accessed August 8, 2011).

Gehm, Ryan. 2011. "Big Losers." *Automotive Engineering International Online* (April 19, 2011):12-18. SAE International, Warrendale, PA.

Gromer, Cliff. 1992. "Ultracar." *Popular Mechanics* no. (April 1992):33-35, 125.

"Keynote: Composites automation evolving necessity." 2010. *Composites World* no. April 12, 2010 (http://www.compositesworld.com/news/keynote-composites-automation-evolving-necessity).

Koganti, Rama and Paul Bennett. 2005. "Carbon Fiber and Super Plastic Aluminum Formed Panel Decklid Manufacturing Development for the Ford GT." SPE ACCE, Troy, MI (http://speautomotive.com/aca).

Lamborghini. 2011. "Lamborghini Model Range." (http://www.lamborghini.com/ Accessed August 8, 2011).

Lexus. "LFA, Product Info, Tech Features." (http://www.lexus-lfa.com/ Accessed August 8, 2011).

———. "Milestones, Bare Chassis Completed." (http://www.lexus-lfa.com/ Accessed August 8, 2011).

———. "Milestones, LFA Line-Off." (http://www.lexus-lfa.com/ Accessed August 8, 2011).

———. "Milestones, Sugo Test Drive." (http://www.lexus-lfa.com/ Accessed August 8, 2011).

Lotus. 2010. "Lotus Cars. Evora Carbon Concept Makes Motorshow Debut." *Car News September 2010* (http://www.lotuscars.com/news/en/evora-carbon-concept-makes-motorshow-debut Accessed August 8, 2011).

"Lotus Elan Again Present At the LA Auto Show." 2011. *News Technology Automotive* no. posted April 16, 2011 (http://www.newstechnologyautomotive.com/lotus-elan-again-present-at-the-la-auto-show Accessed August 8, 2011).

Lutsey, Nicholas. 2010. "Review of technical literature and trends related to automobile mass-reduction technology." *Prepared for the California Air Resources Board under Agreement Number 08-626.* no. UCD-ITS-RR-10-10 (http://www.arb.ca.gov/msprog/levprog/leviii/meetings/051810/2010_ucd-its-rr-10-10.pdf Accessed October 7, 2011).

Masini, Attilio, Andrea Bonfatti, and Paolo Feraboli. 2006. "Carbon Fiber Composites for Improved Performance of the Murcielago Roadster." *Automotive Composites Consortium & Exhibition* (http://speautomotive.com/aca).

Masini, Attilio and Paolo Feraboli. 2003. "Carbon/Epoxy Composites for the Lamborghini Murcielago." *Automotive Composites Consortium & Exhibition* (http://speautomotive.com/aca).

Mayersohn, Norman. 2011. "McLaren democratizing the stuff of 'supercars'; Use of carbon fiber may trickle down to broader market " *New York Times News Service* no. April 15, 2011 (http://www.bendbulletin.com/article/20110415/NEWS0107/104150369/).

"Mercedes-Benz History: Mercedes-Benz Super Sports Cars From AMG." 2011. (http://www.emercedesbenz.com/autos/mercedes-benz/corporate-news/mercedes-benz-history-mercedes-benz-super-sports-cars-from-amg/ Accessed 2011).

"Michigan grant may accelerate low-cost carbon fiber." 2010. *Composites Technology* no. April 1, 2010 (http://www.compositesworld.com/news/michigan-grant-may-accelerate-low-cost-carbon-fiber Accessed August 14, 2011).

Nelson, Jared. 2005. "Corvette Z06 adds carbon fiber fenders." *Composites World* no. High-Performance Composites November 2005 (http://www.compositesworld.com/articles/corvette-z06-adds-carbon-fiber-fenders Accessed August 9, 2011).

Porsche. 2003. Porsche Press Release, "Porsche Carrera GT Wins Popular Science Magazine's Best of What's New Award." no. November 7, 2003 (http://press.porsche.com/news/release.php?id=194 Accessed August 8, 2011).

Robinson, Aaron. 2001. "The Green Brigade - Feature." *Car and Driver* no. May 2001 (http://www.caranddriver.com/features/01q2/the_green_brigade-feature/gm_precept_page_4 Accessed August 9, 2011).

Seabrook, Jeff. 2005. "World Debut of the new BMW M6 - SMG Transmission / Stability." *BMW AG News - BMW at the 75th Geneva Auto Show* no. March 2, 2005 (http://bmwsport.net/index.php?option=com_content&view=article&id=281%3A world-debut-of-the-new-bmw-m6&catid=112%3Abmw-at-the-75th-geneva-auto- show&Itemid=112&limitstart=1).

Sedgwick, David. 2011. "New techniques cut cost of carbon fiber." Automotive News, p. 10, July 2011.

Sudharsan, Ranjani. 2011. "Properties of Carbon Fibre Composites for Automotive Crash Structures." *AutoCRC* no. The "light weighting low down" Automotive Seminar (http://www.autocrc.com/files/File/2011/Properties%20of%20Carbon%20Fibre%20 Composites%20for%20Automotive%20Crash%20Structures.pdf Accessed October 7, 2011).

"VCAMM News. World First Research Facility for Carbon Fibre Innovation." 2010. no. February 6, 2010 (http://www.vcamm.com.au/newsitem.asp?newsID=30 Accessed October 9, 2011).

"Voith to Manufacture Carbon Fiber Parts for Audi." 2011. *autoevolution* no. February 16, 2011 (http://www.autoevolution.com/news/voith-to-manufacture-carbon-fiber- parts-for-audi-31372.html).

"VW buys into BMW's carbon-fibre dream." 2011. *The Economist* no. economist.com March 1, 2011 (http://www.economist.com/node/21016439 Accessed August 8, 2011).

Chapter Two

Carbon Fiber Composite Constituents: Fiber and Resin Types for Automotive Applications

Whatever the medium, there is the difficulty, challenge, fascination, and often productive clumsiness of learning a new method: the wonderful puzzles and problems of translating with new materials.
—Helen Frankenthaler

2.1 Carbon Fiber Primer

Carbon fiber has exceptional mechanical and physical properties that are attractive to the automotive industry, including high strength- and stiffness-to-weight ratios that surpass all other materials used in automotive design, including advanced high-strength steel, magnesium, and aluminum (Fig. 2.1).

Each carbon strand, which is called a tow in the carbon fiber industry, is a bundle of many thousand carbon filaments. Each single filament is a thin tube with a diameter of 5-8 μm and consists almost exclusively of carbon. The atomic structure of carbon fiber consists of sheets of carbon atoms (graphene sheets) arranged in a regular hexagonal pattern. The carbon layers, though not necessarily flat, tend to be parallel to the fiber axis. This crystallographic orientation is called fiber texture. The stronger the fiber texture, i.e., the greater the degree of alignment parallel to the fiber axis, the greater the carbon content and density, as well as tensile modulus, and electrical and thermal conductivities in the fiber direction. Concomitantly, the thermal expansion coefficient and internal shear strength will be lower. Depending upon the precursor to make the fiber, carbon fiber may be turbostratic or graphitic, or have a hybrid structure with both graphitic and turbostratic structures. In graphitic carbon fiber, the layers are stacked in an AB sequence such that half the carbon atoms have atoms directly above and below them in adjacent layers. In turbostratic carbon fiber, the layers of carbon atoms are still well developed and parallel to one another, but are not stacked in any particular sequence, being haphazardly folded, or crumpled, together. (Carbon fibers can also be amorphous, in which the carbon layers, still well developed, are not parallel to one another. An example is pitch-based XN-05 by Nippon Graphite Fiber Corporation.) Carbon fibers derived from polyacrylonitrile (PAN) are turbostratic, whereas carbon fibers derived from mesophase pitch are graphitic after heat treatment at temperatures exceeding 2200°C. Turbostratic carbon fibers tend to

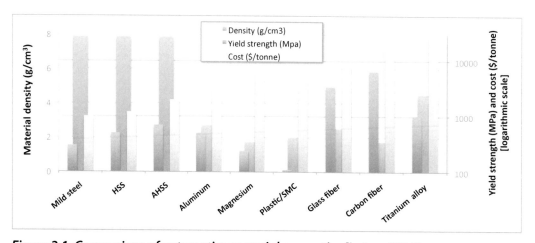

Figure 2.1. Comparison of automotive material properties [Lutsey 2010].
(Source: UCD-ITS-RR-10-10 Report, p. 9)

have high tensile strength, whereas heat-treated mesophase-pitch-derived carbon fibers have high Young's modulus and high thermal conductivity.

Carbon fiber filaments are produced from a polymeric precursor, commonly PAN, rayon, or oil/coal pitch. If the precursor is PAN or rayon (i.e., a synthetic polymer), then the precursor is first spun into filaments using chemical and mechanical processes to align the atoms in a manner that will enhance the mechanical and physical properties of the final carbon fiber. The precursor compositions and the processes used during spinning vary among manufacturers. After spinning or drawing, a common method is to expose the spun fibers to heated air (approximately 300°C) to oxidize the precursor fiber; the oxidized precursor fiber is then carbonized through exposure to progressively higher temperature in an inert environment of gas such as nitrogen or argon (i.e., oxygen-free). The final carbonization occurs at more than 1000°C (more than 2200°C for mesophase pitch-based) to form single columnar filaments with high carbon content (>90%) and establish its strength, modulus, electrical, and other material properties. Post-carbonization heat treatments can further enhance properties. Carbon fiber that was carbonized in the range of 1000–2000°C exhibits the highest tensile strength (approximately 5.5 GPa), while that carbonized in the range of 2500–3000°C exhibits higher Young's modulus (approximately 530 GPa). The carbon fiber is then surface treated or etched to create an effective bonding surface, and a sizing (polymer coating) is usually applied to promote fiber handling, wet-out, and bonding with the resin. Continuous fiber is wound onto spools for use by prepreggers, weavers, braiders, filament winders, and pultruders. The fiber can also be chopped, milled, or pelletized for use in other composite manufacturing processes. Figure 2.2 schematically illustrates the process of producing carbon fiber, and Fig. 2.3 shows spooling of the fiber at the laboratory scale.

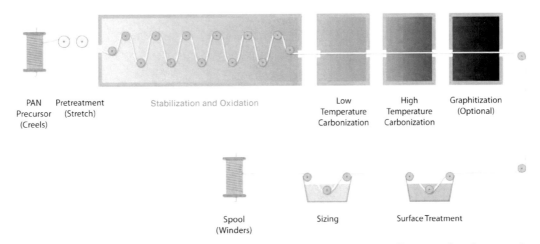

PAN Precursor (Creels)   Pretreatment (Stretch)   Stabilization and Oxidation   Low Temperature Carbonization   High Temperature Carbonization   Graphitization (Optional)

Spool (Winders)   Sizing   Surface Treatment

Figure 2.2. Representative production process of PAN-based carbon fiber [Paulauskas 2011].
(Courtesy of Department of Energy)

*Figure 2.3. Lab scale spooling of carbon fiber tow
[Transportation 2011].*
(Courtesy of Department of Energy)

Carbon fiber yarn is rated by the linear density or the number of filaments per yarn. Linear density is weight per unit length, with 1 g/1000 m defined as 1 tex. The number of filaments per yarn or tow is counted in terms of 1000s or Ks; a 12K tow thus contains 12,000 filaments.

2.1.1 Carbon Fiber History

The development of carbon fibers can be segmented into a few distinct periods: before 1960, between 1960 and 1985, between 1985 and 1991, and after 1991. The year 1960 saw the production of PAN-based carbon fibers with their impressive strength and stiffness. In 1985 was the discovery of a 60-atom carbon cluster forming a spherical closed cage, which was named the buckminsterfullerene. Other fullerenes — cages with different sizes or multi-walled cages — were later found. The year 1991 saw the reporting of carbon nanotubes, and the reporting of single-walled carbon nanotubes followed within a couple years.

During World War II, Union Carbide Corporation investigated the carbonization of rayon and PAN. In 1958, Roger Bacon at Union Carbide created carbon fiber by carburizing strands of rayon, but the process was inefficient and the fibers had low strength and modulus due to a low 20% carbon content. In 1959 and 1962, two processes for manufacturing high-strength and high-modulus carbon fibers from rayon and PAN precursor fibers were invented simultaneously. The carbon fiber developed from PAN precursor by Dr. Akio Shindo at the Agency of Industrial Science and Technology of Japan contained about 55% carbon. In 1963, high-modulus carbon fibers made from pitch were invented. During the early years, several precursor materials were considered for producing carbon fibers: rayon, PAN, pitch, polyesters, polyamides, polyvinyl alcohol, polyvinylidene, poly-phenylene, and phenolic resins. In the early 1970s, large-scale commercial production of PAN-based and isotropic pitch-based carbon fibers began. Anisotropic pitch-based carbon fibers entered the commercial market in the late 1980s. Carbon fibers produced from rayon, PAN, and pitch have been found to offer the greatest mechanical properties, and today are the three major classifications of carbon fiber in industrial production. Currently, more than 90% of the commercially marketed carbon fibers are made from PAN. PAN-based carbon fibers possess higher strength-modulus and failure strain with high yield compared to fibers made from pitch or rayon. Rayon-based carbon fibers have very low yield (10%–30% depending on the precursor) and are difficult to process compared to PAN-based carbon fibers. Pitch-based carbon fibers are used in specialty applications because they provide properties not readily obtained with PAN-based fibers such as low modulus to ultrahigh modulus, and good thermal and electrical conductivity.

Prior to 1991, the carbon fiber industry was driven by the Department of Defense and defense industries. These high-performance carbon fibers are produced in small tows ranging from 1000 (1K) filaments to 12,000 (12K) filaments per tow. After the end of the Cold War and with the collapse of the USSR, the demand for carbon fiber dropped sharply and the carbon fiber industry consolidated and restructured through various acquisitions, similar to contractions in the defense industries. Cytec purchased Amoco's carbon fiber business that was acquired from BASF's manufacturing plant and equipment in Rock Hill, South Carolina, in 1993. Hexcel acquired Hercules PAN and carbon fiber business in Decatur, Alabama, and Salt Lake City, Utah, in 1993. Zoltek bought equipment from Stackpole, Courtaulds, and Cydeca, Mexico. RK Carbon was sold to Sigri Great Lakes Carbon (SGL). Because of the reduced demand for carbon fiber from 1995 through 2000, all carbon fiber producers were operating below their capacity (less than 60%); the price of carbon fiber dropped, as did the dependence on defense and aerospace markets while the market for carbon fiber expanded into commercial aerospace, industrial, sporting equipment, energy, and automotive applications. By 2002, the demand for carbon fiber was increasing, and there became a strong need for carbon fiber at affordable prices to be competitive in these new market segments. As a result, there was a large expansion of carbon producers' manufacturing capacities. Splittable PAN fiber large tow, based on textile industry tow of more than 320K filaments, provided the precursor for producing commercial-grade carbon fiber that met the industrial applications for carbon fiber at affordable and competitive prices. The tow sizes for aerospace-grade carbon fiber range from 1K–12K filaments, and their prices could be as high as $200/lb depending on the carbon fiber class (modulus, strength). Composites manufacturing technologies during the last 50 years were developed and implemented for these small tow sizes. The commercial large-tow sizes are 50K–84K and their current prices are around $8/lb–$12/lb depending on the volume. Large-tow carbon fibers are manufactured in the standard modulus type (approximately 30–35 Msi tensile modulus). Also, large-tow carbon fiber has been successfully used in many industrial applications, such as automotive, sporting equipment, and wind energy. There are some difficulties and issues in the manufacturing of composite structures and components made from large-tow carbon fiber tows using technologies for the small carbon fiber tows. These issues are being addressed and researched to take advantage of the large carbon fiber tow competitive prices. In the last five years, many producers are marketing carbon fiber in 24K, 36K, and 50K tow sizes. These intermediate size tows have been successfully used in weaving and filament winding processing technologies to overcome the large-tow processing difficulties. The PAN-based carbon fiber commercial-grade large-tow producers are Zoltek, SGL Carbon, TohoTenax (at the former Fortafil location), and Mitsubishi/Grafil. It is estimated that current uses of large tow represent approximately 25%–30% of the carbon fiber market; it is expected to increase to approximately 40% by 2012. In the half-century since PAN-based carbon fiber was invented, Asian, U.S., and European companies have successfully manufactured and commercialized PAN-based and pitch-based carbon fibers. Currently, the world's leading carbon fiber producers are: Cytec Industries, Hexcel Corporation, Mitsubishi Rayon Co., Ltd., SGL Group, Toho Tenax (Teijin Limited), Toray Industries, and Zoltek Companies.

Taking Toray as an example, one can see a broad-brush view of the commercial developments of carbon fiber over the decades. In 1982, Toray T300 was used in components of the Boeing B757, B767, Airbus A310, and in the space shuttle Columbia (in the cargo room door). In 1990, Torayca prepreg P2302-19 — which is a high-strength intermediate modulus yarn T800H in toughened epoxy resin 3900-2 — was qualified to the Boeing material specification BMS 8-276 for the Boeing 777. In 2002, Airbus announced that Toray's T800S would be used in the A380 program.

2.1.2 Carbon Fiber Developments

There is one main driver to carbon fiber development as far as the automotive industry is concerned: cost. The mechanical and physical properties of carbon fiber are attractive to the automotive industry, particularly in the current drive toward reducing vehicle weight. However, the cost is too high for use in mainstream vehicles. Carbon fiber produced from the traditional PAN currently costs roughly $24/kg ($10–$12/lb). This cost is almost equally split between precursor and processing costs, and new technology developments are attacking both.

The automotive industry is also pushing for developments in producing carbon fiber with targeted mechanical properties for the end composite. Such targeted properties include strain to failure, compression strength, conductivity, and shear strength. In terms of national interest, there is also a strong desire to develop technology for alternative non-petroleum based precursors that will reduce the carbon fiber's dependency on petroleum resources. Attaining lower cost is also a requirement for non-petroleum based precursors, and hence they are included in the section, Low-Cost Carbon Fiber.

2.1.2.1 Low-Cost Carbon Fiber: Including Non-PAN- and Non-Petroleum-Based Precursors

For CFCs to be implemented into a greater number of applications for future vehicles, the cost of carbon fiber must come down to about half its current cost. This goal has necessitated development of alternative precursors and more efficient production methods, discussed in this chapter, as well as new composite manufacturing approaches, which are discussed in Chapter 4.

Aerospace-grade carbon fibers are produced with specialty PAN-based precursors. Production of these very high quality proprietary fibers requires precise control on polymer composition, purity, molecular weight distribution, and molecular orientation in the fibers. Certain processes are used during the initial conversion steps to retain specific fiber morphological characteristics that are so critical to achieving the desired properties of the resultant fibers. Because the material property requirements for automotive applications are less demanding than those for aerospace, the automotive industry can take advantage of cost savings that less demanding precursor quality and carbon fiber performance will bring.

Across the globe, there are numerous research and development activities underway in academia, industry, and government institutions that are focused on lowering the cost of carbon fiber. The automotive industry has been directly involved in some of these either as individual companies with or without academic or government collaboration, or through the Automotive Composites Consortium (ACC) that is part of the U.S. Automotive Materials Partnership (USAMP). USAMP is an industry partnership between Chrysler Group LLC, Ford Motor Company, and General Motors Company that was established in 1993 to develop materials and processes for high-volume production of vehicles that will lead to vehicles that are lower weight, more recyclable, and affordable, without detracting from quality or durability. USAMP is part of the United States Council for Automotive Research (USCAR) that was established in 1992. The ACC has been working with Oak Ridge National Laboratory (ORNL) and its partners to develop technologies that would enable the production of commercial-grade carbon fiber at $5/lb–$7/lb. This work is being conducted with Department of Energy (DOE) sponsorship. (Some reports indicate the cost must actually reach $3–$5/lb for the automotive industry to benefit from carbon fiber technology [Griffith, Compere, and Leitten 2004].)

Lower-cost alternative precursors under investigation include textile-grade PAN fibers, polypropylene (PP) fibers, lignin-based feedstocks, and ethanol fuel [ORNL 2009]. Progress made in these areas over recent years is briefly summarized below.

Textile-Grade PAN-Based CF

PAN-based acrylic textiles are lower-cost, higher-volume fibers compared to the aerospace and commercial specialty PAN-based carbon fiber precursors. A cost model published in 2007 showed that textile-based precursors could reduce the cost of carbon fiber by more than $4.40/kg. This potential cost reduction is due to the high volumes of the textile industry, and the higher oxygen content of the precursor will permit faster oxidation times. Textile precursors would be relatively easy to incorporate in existing carbon fiber facilities because the processing of textile precursors is similar to that of conventional carbon fiber.

In the early 2000s, through a separate contract with the Automotive Lightweighting Materials program, Hexcel Corporation developed the science that enables textile-grade PAN to be used as a carbon fiber precursor [Warren et al. 2008]. That science included chemically pretreating the textile precursor while the fiber was still uncollapsed. Hexcel was able to produce satisfactory samples but only with one specific textile fiber and under one set of processing conditions [Baker, Gallego, and Baker 2009]. Under the FreedomCAR initiative of DOE, low-cost carbon fibers are being produced at ORNL from chemically modified textile PAN fibers in collaboration with FISIPE S. A. of Portugal. Two different textile-grade PAN-based precursors have been developed [Warren et al. 2008]. Fibers were produced and chemically pretreated in Portugal and then sent to ORNL for evaluation, oxidative stabilization, and carbonization. Single filaments and 26K filament tows have been tested, with the tows showing tensile modulus of 200–228 GPa and strength of 2.14–2.69 GPa. The FY 2010 Progress Report for Lightweighting Materials [Energy 2010] reports achieving strengths of 450 ksi for full-scale tows and modulus of 35 Msi. During FY 2010, the team consulted with SGL Group on the applicability of FISIPE precursors in their manufacturing systems and provided small samples to BMW and SGL for evaluation. In December 2010, FISIPE inaugurated its Cogeneration Plant and Pilot Plant for the production of carbon fiber. Textile-grade PAN based precursors are expected to be available soon; FISIPE manufacturing facilities are anticipated to be modified in ORNL fiscal year 2013 [Paulauskas 2011].

Polyolefin-Based CF
Ideally, to obtain a low-cost carbon fiber, one would consider a low-cost precursor that produces high yield. Polyolefins are the least expensive petroleum-base polymers, and they offer high yield, especially compared with the competing precursors lignin and textile-PAN [Warren et al. 2008]. Abdullah [2003] reported that Hexcel, within the Advanced Lightweighting Materials (ALM) Low-Cost Carbon Fiber program, concluded that linear low density polyethylene (LLDPE) and polypropylene (PP) precursors were two of the three most promising low-cost carbon fiber (LCCF) precursors, along with textile-PAN. In 2002, the program demonstrated the conversion of LLDPE to LCCF, meeting its program and cost predictions. However, the issues of sulfuric acid recycling and availability of the LLDPE precursor led to the conclusion that LLDPE-based precursor technology would need more development to compete with the textile-grade PAN-based precursors. Hexcel did not pursue LLDPE as a precursor, but PE-based precursors did continue with ORNL and its partners.

However, another ALM project [ALM-18899] has continued this pursuit of polyolefinic precursors [Baker, Gallego, and Baker 2009]. In 2009, the DOE Lightweighting

Materials program reported the production of polyolefinic precursor fibers of varied compositions and different diameters (< 15 μm), as well as tensile strengths of 10-20 ksi [Baker, Gallego, and Baker 2009]. In 2010, the DOE Lightweighting Materials demonstrated carbonized fiber properties of 150 ksi and 15 Msi tensile strength and modulus, respectively [Energy 2010]. This is on par with the current lignin-based carbon fibers, and efforts indicate that even better mechanical properties can be obtained relatively soon. The goal for 2012 is to demonstrate fiber properties with strength and modulus values greater than 250 ksi and 25 Msi, respectively [Energy 2010].

Lignin-Based CF
Lignins were first evaluated in the 1960s and 1970s in Japan [Warren et al. 2008]. Approximately 10 years ago, ORNL and PNNL began their efforts to continue the work on lignin-based carbon fiber. Lignin-blend feedstocks were selected for detailed investigation based on low cost, high production volume, and ability to be melt-spun and readily stabilized, carbonized, and graphitized. Lignin has the potential to be a significantly cheaper precursor than textile grade PAN, and has the advantage of being largely independent of the petroleum market.

Black liquors from alkaline pulping are a potential source of low-cost lignin in quantities sufficient to replace half of the ferrous metals used in automobiles and light trucks [Griffith, Compere, and Leitten 2004].

Currently, the only source of lignin on the scale envisioned for lower-cost carbon fiber production is Kraft pulping of wood, and MeadWestvaco Corporation is the only commercial producer of Kraft lignin products. Their hardwood lignin, precipitated from pulping liquors, contains significant impurities, such as fibers, salts, and carbohydrates. These contaminants can cause point defects in the carbon fibers. They also plug multifilament spinning dies [Warren et al. 2008]. The contaminants appear to be removed effectively by a combination of prefiltration, fractional precipitation, and washing. These processes are cost-effective and can be readily scaled-up for use in pulp mills. A two-step process for extrusion of lignin-blend feedstocks has been used to decrease offgassing during extrusion, and high-shear spinning dies have been used to improve fiber structure. Hardwood lignin purified by MeadWestvaco's alternative organic method had excellent structural uniformity and was readily melt spinnable; its low melting point (about 130°C) prevented an acceptable rate of fiber stabilization [Warren et al. 2008].

Although not on commercial scale, Kruger Wayagamack of Canada has produced a much cleaner softwood lignin than the Kraft lignin that is currently produced commercially, and in fact did not need purification to be continuously melt spun. However, because of the chemistry of softwood lignin, it does not have well-defined melting points and cannot be melt spun as-produced. ORNL overcame this by adding appropriate plasticizing agents, which include certain hardwood lignins, and carbon fiber filaments 10 μm in diameter were successfully continuously spun. However, the fibers were not as structurally sound as those from the MeadWestvaco purified hardwood lignin. The tensile modulus and strength were only about 70 GPa (10 Msi) and 700 MPa (100 ksi), respectively. Nonetheless, it seems promising that with refinements in the Kruger process and by adding tension during the thermal processing and increasing the heat treatment temperature, it should be possible to produce lignin-based carbon fiber with better mechanical properties without requiring a purification process [Warren et al. 2008].

The FY 2009 Progress Report on Low-Cost Carbon Fiber from the DOE Vehicle Technologies Program [Baker, Gallego, and Baker 2009] reports some notable accomplishments: demonstration of continuous melt spinning of lignin precursor fiber tow from two distinct lignin sources with a filament diameter of 10 μm and without the need for purification, and a spinning speed of 1500 m/min, or 2.5 times the baseline speed. It was also determined that softwood lignin crosslinks more readily than hardwood lignin, and that an 80/20 blend (by weight) of Organosolv-pulped hardwood lignin and Kraft-pulped softwood lignin could be stabilized at ten times the rate of the 100% Organosolv (only) lignin fiber. Unfortunately, it was also discovered that a high degree of nano-sized porosity develops in the carbon fiber as a result of in-situ carbon gasification during the thermal processing. This porosity is detrimental to the mechanical properties of the carbon fiber and is a major deficiency.

Although research and development is continuing in an effort to improve the mechanical properties of lignin-based carbon fiber, it is likely that these current properties are adequate for some automotive CFC applications, and efforts are underway to assess such feasibility. It was reported by Warren et al. [2008] that previous design studies by the ACC showed that 250 ksi (1.72 GPa) strength and 25 Msi (172 GPa) modulus fibers were sufficient to mold as thin a part as is practical for the majority of automotive applications.

Other potential valuable resources for carbon fiber production are the lignin by-products from cellulosic ethanol fuel production in biomass refineries. These by-product lignins are much purer than Kraft lignins [Warren et al. 2008].

Another noteworthy effort underway in the lignin arena is that of Zoltek Companies, Inc. In August 2011, the DOE announced $3.7M funding for Zoltek's work on developing a novel low-cost route to carbon fiber using a lignin/PAN hybrid precursor and carbon fiber conversion technologies leading to high-performance, low-cost carbon fiber.

Pitch-Based Carbon Fiber
In the search for lower-cost precursors for high-performance carbon fiber (HPCF), others have pursued advanced development of pitch-based carbon fibers. The manufacture of PAN-based HPCF requires wet/dry melt spinning processes that involve a costly wet chemical bath and relatively long stabilization time. Pitch, however, can

be spun through melt spinning, which is cheaper than wet/dry spinning [Liu 2010]. Conoco-Philips was the most notable pursuer of this technology during the late 1990s through the early 2000s, and the University of Tennessee Space Institute (UTSI) has continued this development with an approach to the processing that was anticipated to significantly reduce the cost of the end carbon fiber. They started with solvated mesophase pitch, and the preparation process consists of spinning of green (non-stabilized) fibers via spraying melted pitch out of a spinneret-jet and drawing with blown hot air; stabilization by exposing the green fibers to 150–400°C to remove solvents and allow oxygen to diffuse into the fibers and enable cross-linking that strengthen the bonds; carbonization/graphitization at 1000–3000°C in an inert environment for a short period of time to remove all non-carbon elements from the fibers and enable further cross-linking; and surface treatment to change the surface properties to provide better adhesion to the resins. It was anticipated that this process would significantly reduce the cost of the carbon fibers because the solvated pitch has a lower soft point/melting point, making it easier to spin. The air blowing melt spinning is a high-volume production fiber spinning process that is much cheaper than conventional melt spinning, and the spun pitch fiber can be stabilized in a very short time. However, the melt-blowing process, which created turbulence and micro-vortices, produces fibers that are kinky. Without excellent alignment the carbon fibers could not attain the required high strength and modulus [Liu 2010].

The current commercially produced pitch-based carbon fibers can achieve much higher modulus than PAN-based fibers, and in some cases higher toughness, but unfortunately there is currently no cost advantage of pitch-based over PAN-based.

2.1.2.2 Targeted Carbon Fiber Properties

Two of the most sought after improvements in mechanical performance of CFCs are higher strain to failure and higher shear strength.

In 2010, the DOE Lightweighting Materials program reports using ORNL's advanced surface treatment on single-tow carbon fiber, and mechanical testing of the ensuing composite material showed an increase in short beam shear strength from 6 ksi to 10 ksi [Energy 2010].

N12, an upstart company working with nanostitching and fuzzy fiber technologies spun off from MIT, also has a method for increasing the interlaminar shear strength of CFCs. The technology involves using aligned carbon nanotubes (CNTs) to reinforce and tailor existing composite constructions. In nanostitching, CNTs that are aligned perpendicular to the carbon fiber act to stitch together adjacent layers of carbon fiber. In fuzzy fibers, the CNTs protrude off the carbon fibers, creating what appears to be a fuzzy carbon fiber, with the CNT fuzz increasing adhesion to the matrix. In the laboratory, fuzzy fiber composites have demonstrated a 70% increase in interlaminar shear strength (short beam shear) [Wardle 2010].

Several researchers are working on carbon nanotubes or vapor-grown carbon nanofibers. Work at University of Dayton Research Institute centered around blending modest amounts of vapor-grown carbon nanofiber (VGNF) into epoxy resin [Klosterman et al. 2007; Rojas, Maruyama, and Barrera 2006]. The goal is to incorporate the nanotubes or nanofibers within the carbon fiber composite to increase strength, modulus, and thermal and electrical conductivities. The methodologies have varied, and the level of success in terms of mechanical property improvement has varied with them. The improvement in thermal and electrical conductivities has been more significant and more consistent. However, these nano incorporations do show promise for eventually providing significant improvements in mechanical performance.

Attempts to increase the strain to failure of carbon fiber and its composites include amorphous mesophase pitch-based carbon fiber (e.g., XN05 by Nippon Graphite). Efforts are also underway within the ORNL lignin-based carbon fiber program to create lower-modulus fibers that retain the current 150 ksi strength, thereby providing a higher strain to failure.

In a different vein, TorTech Nano Fibers Ltd., a joint venture formed in 2011 between Plasan and Q-Flo, is developing a carbon nanotube-based yarn, which can be woven into the strongest manmade material that will still be flexible and lightweight. This could lead to a breakthrough in structural composites and lightweight armor.

2.1.3 Developments in Conversion Processes and Post-Treatments

Stabilization and Oxidation

Converting precursor fibers to finished carbon fibers consumes both a lot of time and energy, and is thus expensive. In conventional CF conversion, the rate-limiting processes are thermal stabilization and oxidation, which occur in three to four successive furnaces at temperatures around 200–250°C. Conventional oxidation accounts for over 80% of the conversion time, so if these steps are accelerated the carbon fiber conversion cost would be reduced.

In work done at ORNL, electron beam, ultraviolet, and thermo-chemical methods were evaluated, and a thermo-chemical based plasma was found to be the most acceptable PAN precursor stabilizer method. ORNL's Advanced Stabilization process would replace the first of four oxidative stabilization ovens used in conventional processing, and an Advanced Oxidation process would replace those final three ovens [Warren et al. 2008]. These advanced processes have been developed with conventional carbon fiber precursors, and have shown a significant reduction in residence time in the thermal oxidative stabilization ovens, from 90–120 minutes to less than 35 minutes [Warren et al. 2008].

Carbonization/Graphitization

ORNL has developed a Microwave Assisted Plasma (MAP) technology that would replace the low- and high-temperature carbonization ovens in conventional processing.

Microwave/plasma processing of stabilized polyacrylonitrile (PAN) feedstock to pro-duce industrial-grade carbon fibers has moved toward pilot scale with fiber properties comparable to those of commercial materials. The microwave/plasma process can also be used to produce industrial-grade carbon fiber from chemically treated textile PAN [Griffith, Compere, and Leitten 2004]. To carbonize and graphitize large tow (>48k) commercial-grade CF, the MAP process has several advantages over the conventional thermal process, showing reductions in several areas: residence time, capital invest-ment, energy demand, operation temperatures, equipment start-up and shut-down times, and hazardous emissions and/or emission treatment requirements. Additionally, the MAP process provides a controllable surface chemistry, which leads to improved composite properties [Warren et al. 2008].

2.1.4 Future Developments around the Carbon Fiber

Near-term research efforts in the conversion process include advanced surface treat-ment and sizing, plasma modification of the fiber surface topology to enable me-chanical interlocking with the resin, and modeling the oxidation kinetics. Near-term research efforts in other non-conversion processes include tow splitting, in which large tows are processed and then split for use as small tows, developing non-spooled product forms that are amenable to high-volume industries such as automotive, and carbon fiber recycling [Warren 2011]. (Recycling of carbon fiber is discussed in Chapter 6.)

2.2 Resins for the Automotive Industry

Traditionally, CFCs adopted from the aerospace industry for use in the automotive industry are fabricated with an epoxy resin. Epoxy is a thermoset, and other thermoset resins that are often used in CFCs include bismaleimide (BMI), phenolic resins, poly-urethane (PU), unsaturated polyester (PET), vinyl ester, and polyimide resins. The ep-oxies, unsaturated polyesters, and vinyl esters are currently the most commonly used thermosetting resins. Thermoset resins cure permanently by irreversible cross linking at elevated temperatures. This characteristic makes the thermoset resin composites very desirable for structural applications.

Carbon fiber can also be incorporated in most thermoplastic resins, including poly-propylene (PP), polyamide (PA), acetal, ABS, polyethersulfone (PES), polyetherimide (PEI), polyetheretherketone (PEEK), polyetherketone-ketone (PEKK), and polyphen-ylenesulfide (PPS). Thermoplastic resins remain solid at room temperature. They melt when heated and re-solidify [Warren et al. 2008] when cooled. The long-chain polymers do not chemically cross link. Thermoplastic resins are beginning to see more implementation into semi-structural and structural applications, not only in the auto-motive industry but in the aerospace, marine, and military industries as well.

2.2.1 Characteristics of Different Resins Used in CFCs

Unsaturated Polyesters

The unsaturated polyester accounts for about 75% of all polyester resins used in the U.S.A. It is produced by the condensation polymerization of dicarboxylic acids and dihydric alcohols. The formulation contains an unsaturated material such as maleic anhydride or fumaric acid, which is a part of the dicarboxylic acid component. The formulation affects the viscosity, reactivity, resiliency, and heat deflection temperature (HDT). The viscosity controls the speed and degree of wet-out (saturation) of the fibers. The reactivity affects cure time and peak exotherm (heat generation) temperatures. High exotherm is needed for a thin section curing at room temperature, and low exotherm is needed for a thick section. Resilient or flexible-grade composites have a higher elongation, lower modulus, and HDT. The HDT is a short-term thermal property which measures the thermal sensitivity and stability of the resins. The main advantages of unsaturated polyester are its dimensional stability and affordable cost. Other advantages include ease in handling, processing, and fabricating. Some of the special formulations are high-corrosion resistant and fire retardant. This resin is one of the best values for a balance between performance and structural capabilities.

Epoxies

The epoxies used in composites are mainly the glycidyl ethers and amines. The material properties and cure rates can be formulated to meet the required performance. The largest use of carbon fiber/epoxy composites is in aerospace, but they are also generally found in marine, automotive, electrical, and appliance applications. The high viscosity in epoxy resins limits their use to certain processes such as molding, filament winding, and hand lay-up. The proper curing agent should be carefully selected because it will affect the type of chemical reaction, pot life, and final material properties. Although epoxies can be expensive, they may be worth the cost when high performance is required.

Vinyl Esters

The vinyl ester resins were developed to provide the workability of epoxy resins and the fast curing of polyesters. Vinyl esters have higher physical properties than polyesters and cost less than epoxies. The acrylic esters are dissolved in a styrene monomer to produce vinyl ester resins which are cured with organic peroxides. A composite product containing a vinyl ester resin offers high toughness and excellent corrosion resistance. However, vinyl ester resin can shrink as much as 10% by volume during curing, compared to 3–4% for epoxies [Xu, Mase, and Drzal 2003].

Polyurethanes

Polyurethanes (PUs) are produced by combining polyisocyanate and polyol in a reaction injection molding process or in a reinforced reaction injection molding process. They are cured into very tough and high-corrosion-resistant materials which are found in many high-performance paint coatings. Because two-part PUs have a fast reaction time and rapid viscosity increase upon mixing, they have traditionally been limited to small parts or continuous processes such as pultrusion.

Phenolics, Bismaleimides, and Polyimides

Phenolics, bismaleimides (BMIs), and polyimides are used in aerospace applications because of their high temperature resistance. Automotive under-hood temperatures can exceed the abilities of standard and some specialty epoxies. If CFCs are desired under hood, one may need to consider one of these high-temperature resins. One would also have to consider changes in processing that these resins may require compared with traditional epoxies.

The phenolic resins are rated for good resistance to high temperature, good thermal stability, and low smoke generation. Polyimides have been used for decades as matrix resins or high-temperature applications that exceed the capability of epoxies (177°C) and bismaleimides (232°C) [Chuang, Criss, and Mintz 2010]. A notable example is the polyimide composite outer bypass duct for the F-404 engine that replaced the titanium duct, leading to a 30% cost savings and 12% weight savings. The F-404 engine is manufactured by GE Aviation and is used in military aircraft worldwide, including the F/A-18 Hornets and F-117 stealth fighters.

Thermoplastics

Thermoplastics in general are easier to process, have greater toughness and high strain to failure, but are limited to operating temperatures much lower than thermosets. In the automotive industry, the two most common thermoplastics currently used in fiber reinforced composites (primarily glass) are polypropylene and nylon. There are many different formulations and/or copolymers that fall within these two families of resins. Different formulations target certain mechanical, chemical, or processing characteristics. There are also a number of technologies that enhance the compatibility between the fiber and resin. Most of the formulation technology is proprietary, and it is suggested that the interested reader inquire with individual resin suppliers for explanations and data.

2.2.2 Resin Developments

Three main drivers continue to push resin development for the automotive industry: cycle time, performance (mechanical and physical properties), and processing methods. Key developments that have occurred recently or are under investigation are discussed next. Because epoxies are currently the dominant resin used in CFCs, much development work has been focused on epoxies.

2.2.3 Resins with Reduced Cure Time

An example of a reduced-cure-time epoxy resin is that used in Hexcel's HEXPLY® M77 prepreg. Its M77 epoxy is a modified and toughened epoxy with a glass transition temperature Tg of 120°C (250°F) that can be cured over a temperature range of 80–160°C. The epoxy is advertised to have a two-minute cure time at 150°C, although the ideal cure cycle is stated to be seven minutes at 120°C.

Hunstman has developed a polyurethane-based resin for use with glass and carbon fiber reinforcement that has a proprietary catalyst system that permits control over the viscosity-time profile, and can achieve cure in less than five minutes or as long as several hours [Stephenson 2011].

Hexion developed basic technologies for very fast cure epoxy with reduced exotherm combined with excellent fiber wetting. As of 2010, their resin transfer molded (RTM) cycle time for a composite part weighing several kilograms was in the range of 4–5 minutes [Reichwein et al. 2010].

2.2.4 Resins with Targeted Performance

To increase the toughness of carbon fiber/epoxy composites, Toray adds thermoplastic particles that toughen the interlayer regions and suppress crack propagation. Toray's high-toughness T800H/3900-2 incorporates this technology, and the material was the first composite qualified for primary structures of Boeing's civil transport and was implemented in its 777 [Toray 2011].

Epoxy resins typically contain multifunctional epoxies as all or part of the resin system to attain the high crosslink densities that are needed to achieve a high T_g. However, this highly cross-linked network is brittle and difficult to toughen. Huntsman has developed a new epoxy (MY816) using Naphthalene epoxy resin to attain a high T_g with a lower degree of cross-linking. This creates an epoxy matrix system which is inherently tougher and is also much easier to toughen further with various toughening materials [Hoge, Hoegy, and Corbett 2010].

He and Li [2010] investigated the effect that different loadings of novolac resin (NR) has on the mechanical properties of carbon fiber/epoxy composites. Evaluating shear and impact performance, they found that composites containing up to 13% NR exhibited higher shear strength, higher elongation, and higher energy to break.

Although carbon fiber/epoxy composites are quite common, they do have some other drawbacks besides low toughness. These include: a relatively low Tg and instability around 170°C which limits their high-temperature use; in-use microcracking due to different thermal expansion coefficients of the fibers and resin which degrades mechanical performance; and the required clean, refrigerated storage of prepregs that have a relatively short shelf life [Bender and Economy 2008]. These researchers developed a family of aromatic thermosetting copolyesters (ATSP) that can be formed and completely cured as individual lamina, thus needing no clean refrigerated storage and having a long shelf life. These cured lamina can be made into thick section composites through interchain transesterification reactions at the lamina interfaces. Results indicate the carbon/ATSP composite has tensile and bending performance comparable to carbon/epoxy but has a Tg over 200°C. The ATSP also has a very low 0.3% moisture uptake compared to 2% or more for epoxies [Bender and Economy 2008]. Further work [Samad and Economy 2009] demonstrated that ATSP that is tailored to have a liquid crystalline structure results in reduced stresses at the fiber/matrix interface and better thermal fatigue resistance compared to epoxy. This liquid crystalline polymer can match the thermal expansion coefficient of carbon fiber, thereby minimizing residual stresses [Parkar and Economy 2010]. It was also found that the presence of liquid crystalline character improves fracture toughness. ATSP is stable in air up to 350°C, making it a candidate for automotive under hood applications [Parkar and Economy 2010].

Creating a composite of carbon fiber reinforced vinyl ester requires better fiber-matrix adhesion. Work by Xu, Mase, and Drzal [2003] found the volume shrinkage of vinyl ester during cure could create thermal residual interfacial tensile stress which would decrease the adhesion between the fibers and matrix. A specially formulated epoxy sizing that swells in vinyl ester could counteract the cure volume shrinkage of the matrix.

2.2.5 Resins for Targeted Processing Methods

As mentioned earlier, polyimides find use in high-temperature applications that are beyond the capabilities of epoxies (177°C) and BMIs (232°C). In recent years, efforts have been devoted to developing solvent-free, low-melt-viscosity (10-30 poise) imide resins that can be resin transfer molded as opposed to the conventional fabrication from prepregs [Chuang, Criss, and Mintz 2010]. Chuang et al. have demonstrated a new method to produce high-Tg, low-melt-viscosity imide resins without the use of solvent, that can be processed in cost-effective manufacturing methods such as RTM and resin infusion [Chuang et al. 2007; Chuang, Criss, and Mintz 2010]. Composites fabricated via high-temperature RTM exhibited outstanding high-temperature performance and property retention, including open-hole compression strength and short beam shear [Chuang, Criss, and Mintz 2010].

Targeting vacuum assisted resin transfer molding (VARTM), Depase et al. developed preformed particle modified tackifiers to create interlayer toughened composites [Depase, Hayes, and Seferis 2001]. Although VARTM is a promising low-cost alternative to autoclave processing, the mechanical performance of composites made from VARTM (and conventional RTM) is inferior to those of prepregged composites. Epoxy-based tackifiers were modified with preformed rubber particles and sprayed onto the carbon fabric. The toughened tackifiers did not affect the VARTM process, and the resulting composite exhibited over 70% increase in mode I and mode II interlaminar fracture toughness, with a slight decrease in interlaminar shear strength, compared with the conventional VARTM composite [Depase, Hayes, and Seferis 2001].

In 2010, Mihalich reported on a multi-year program developing an advanced composite that is tough, durable, emits no volatile organic compounds (VOCs) during production or in use, has good mechanical properties, permits high fiber volume fractions, is low cost, and is fully recyclable [Mihalich 2010]. Although the program used fiberglass for reinforcement, the resin and manufacturing process are equally amenable to carbon fiber. Cyclic oligomers of polybutylene terephthalate (cPBT) and fiberglass prepregs were made and evaluated in vacuum bag (VB), vacuum infusion (VI), and VARTM. These processing methods are traditionally used with thermoset resins, not thermoplastic resins, which cPBT is [Mihalich 2010]. The cPBT easily impregnates fibrous reinforcements because of its low melt temperature (<200°C) and very low initial melt viscosity (<40cP). This allows fiber volume fractions of 50% to be easily achieved, and 60% is not uncommon. Using VB consolidation, cPBT prepreg was manufactured into a prototype trailer bed 13.6 m long, 2.5 m wide, and 8–15 mm thick. At the time, it was the largest single thermoplastic structural component ever produced.

Hunstman Polyurethanes has developed a polyurethane-based resin system whose pot life and cure can be controlled by the processors [Stephenson 2011]. This new resin, called VITROX, combines isocyanates, polyols, and a unique proprietary catalyst that provided processing control that is otherwise unachievable with PURs. VITROX has the potential to be used in RTM, vacuum-assisted infusion, filament winding, and other processes, with significantly shorter cycle times than competitive resins. In the cured state, the composites are tough, impact resistant, and have mechanical properties that can exceed those of some epoxies. However, the possible generation of voids from the creation of carbon dioxide during cure must be watched.

There is a significant desire in both the automotive and aerospace communities to develop and implement carbon fiber reinforced thermoplastics. Motivating factors include lower-cost materials, lower-cost composite manufacturing, more options for composite manufacturing, and recyclability. In June 2011, TenCate Advanced Composites and Toray Industries Inc. announced a long-term agreement for producing carbon fiber thermoplastic composites [TenCate 2011]. Their goal is to support the growing demand for thermoplastic prepreg in the aerospace industry, and the anticipated demand in the automotive industry.

2.3 Closing Thoughts

The main factors in the automotive industry driving fiber development and resin development center around cost, performance, cure time, and processing method. The years 2010 and 2011 have seen an incredible amount of cooperation and partnerships between companies operating at different points in the value stream to bring new materials and processing technologies to market quicker. Fiber and resins, and their developments, have just been discussed in relative isolation. It is important to recognize that these are not independent of the other aspects of manufacturing a carbon fiber composite component, nor are they necessarily independent of each other. Different resins process differently with regard to the time, temperature, and pressure required for fiber wet-out and consolidation. In addition, different fiber-resin constructions require different processing methods. Carbon fiber constructions will be discussed in Chapter 3, and manufacturing processes will be discussed in Chapter 4.

References

Abdullah, Mohamed G. 2003. "Low-Cost Carbon Fiber Development Program, Automotive Lightweighting Materials FY 2003 Progress Report." (http://www1.eere.energy.gov/vehiclesandfuels/pdfs/alm_03/5_low-cost_carbon_fiber.pdf Accessed August 14, 2011).

Baker, F. S., N. C. Gallego, and D. A. Baker. 2009. "Low-Cost Carbon Fiber from Renewable Resources, FY 2009 Progress Report for Lightweighting Materials. " (http://www1.eere.energy.gov/vehiclesandfuels/pdfs/lm_09/7_low-cost_carbon_fiber.pdf Accessed August 14, 2011).

Bender, Samantha E. and James Economy. 2008. "Improved Matrix for Carbon Fiber Composites for Aircraft." *40th International SAMPE Technical Conference - Memphis, TN - Sep 8 - 11, 2008.*

Chuang, Kathy C., Jim M. Criss, and Eric A. Mintz. 2010. "Polyimides Based on Asymmetric Dianhydrides (II) (A-BPDA VS A-BTDA) for Resin Transfer Molding (RTM)." *SAMPE 2010 - Seattle, WA May 17-20, 2010.*

Chuang, Kathy C., Jim M. Criss Junior, Eric A. Mintz, Daniel A. Scheiman, Baochau N. Nguyen, and Linda S. McCorkle. 2007. "Low-Melt Viscosity Polyimide Resins for Resin Transfer Molding (RTM) II." *SAMPE 2007 - Baltimore, MD June 3 - 7, 2007.*

Depase, Edoardo P., Brian S. Hayes, and James C. Seferis. 2001. "Interlayer Toughened VARTM Composites Using Preformed Particle Toughened Tackifiers." *33rd International SAMPE Technical Conference - Seattle, WA - November 5 - 8, 2001.*

Energy, U.S. Department of. 2010. "2010 Annual Progress Report, Lightweighting Material." (http://www1.eere.energy.gov/vehiclesandfuels/pdfs/program/2010_lightweighting_materials.pdf Accessed August 14, 2011).

Griffith, W. L., A. L. Compere, and Junior C. F. Leitten. 2004. "Lignin-Based Carbon Fiber for Transportation Applications." *36th International SAMPE Technical Conference - San Diego, CA - November 15 - 18, 2004.*

He, Hongwei and Kaixii Li. 2010. "Mechanical characterization of carbon-fiber/epoxy composites modified using novolac resin." *Society of Plastics Engineers, Plastics Research Online* no. 10.1002/spepro.003382. doi: 10.1016/j.polymer.2007.11.030

10.1016/S0266-3538(01)00022-7

10.1016/j.polymer.2007.02.021

10.1016/j.compstruct.2006.04.032

10.1016/j.compositesa.2008.09.006

10.1002/app.29085.

Hoge, James E., Norbert Hoegy, and Stephanie A. Corbett. 2010. "New Di-Functional Naphthalene Based High Performance Epoxy Resin." *SAMPE 2010 - Seattle, WA May 17-20, 2010.*

Klosterman, Don, Melissa Williams, Chris Heitkamp, Regina Donaldson, and Charles Browning. 2007. "Fabrication and Evaluation of Epoxy Nanocomposites and Carbon/Epoxy Composite Laminates Containing Oxidized Carbon Nanofibers." *SAMPE 2007 - Baltimore, MD June 3 - 7, 2007.*

Liu, Chang. 2010. "Mesophase Pitch-based Carbon Fiber and Its Composites: Preparation and Characterization." *Masters Thesis, University of Tennessee, 2010* (http://trace.tennessee.edu/utk_gradthes/816).

Lutsey, Nicholas. 2010. "Review of technical literature and trends related to automobile mass-reduction technology." *Prepared for the California Air Resources Board under Agreement Number 08-626.* no. UCD-ITS-RR-10-10 (http://www.arb.ca.gov/msprog/levprog/leviii/meetings/051810/2010_ucd-its-rr-10-10.pdf Accessed October 7, 2011).

Mihalich, James. 2010. "Production of Class-8 Truck Trailer Bed Using cPBT Thermoplastic Prepreg & Vacuum-Bag Processing." SPE ACCE, Troy, MI (http://www.speautomotive.com/aca).

ORNL. 2009. "Low Cost Carbon Fiber Composites Overview and Applications." *Workshop on Low Cost Carbon Fiber Composites for Energy Applications, March 3-4, 2009, Oak Ridge TN* (http://www.ms.ornl.gov/pmc/carbon_fiber09/pdfs/LCCF_Overview_20090225.pdf).

Parkar, Zeba and James Economy. 2010. "Orientational Order Induced by Carbon Fiber in Aromatic Thermosetting Copolyester Matrix." SPE ACCE, Troy, MI (http://www.speautomotive.com/aca).

Paulauskas, Felix, L. 2011. "Development of Low-Cost, High Strength Commercial Textile Precursor (PAN-MA)." (http://www.hydrogen.energy.gov/pdfs/review11/st099_warren_2011_p.pdf Accessed October 8, 2011).

Reichwein, Heinz-Gunter, Paul Langemeier, Tareq Hasson, and Michael Schenzielorz. 2010. "Light, Strong and Economical - Epoxy Fiber-Reinforced Structures for Automotive Mass Production." SPE ACCE, Troy, MI (http://www.speautomotive.com/aca).

Rojas, Grace, Benji Maruyama, and Enrique Barrera. 2006. "CNT/VGCF Reinforced Epoxy/CF Composites: The Role of Nanofibers." *38th International SAMPE Technical Conference - Dallas, Texas - Nov 6-9, 2006.*

Samad, Zeba Farheen Abdul and James Economy. 2009. "Improved Matrix Materials for High-Performance Carbon Fiber Composites Aromatic Thermosetting Copolyester." SPE ACCE, Troy, MI (http://www.speautomotive.com/aca).

Stephenson, Scott. 2011. "A new "tunable" polyurethane could revolutionize composites." *Composites Technology* no. February 2011 (http://www.compositesworld.com/columns/a-new-tunable-polyurethane-could-revolutionize-composites).

"TenCate to use Toray carbon fiber in manufacture of thermoplastic prepregs." June 27, 2011. *Composites World* (http://www.compositesworld.com/news/toray-will-supply-carbon-fiber-to-tencate-for-thermoplastic-prepregs Accessed August 17, 2011).

Toray. 2011. "Composite Materials Research Laboratories _ R&D Centers _ R&D Network _ Research and Development _ TORAY." (http://www.toray.com/technology/network/net_004.html Accessed August 14, 2011).

"Transportation Solutions Using Carbon Fiber." 2011. *Oak Ridge National Laboratory* (http://www.ornl.gov/sci/ees/transportation/pdfs/CarbonFiber_Brochure.pdf Accessed October 7, 2011).

Wardle, Brian L. 2010. "Bulk Nanostructured Materials and Advanced Composites." *Nano-Engineered Composite Aerospace Structures Consortium* no. 2010 Gordon Research Conference on Composites, Ventura, CA (January 20, 2010).

Warren, C. David. 2011. "Low Cost Carbon Fiber Overview." (http://www1.eere.energy.gov/vehiclesandfuels/pdfs/merit_review_2011/lightweight_materials/lm002_warren_2011_o.pdf): Accessed August 14, 2011.

Warren, C. David, Felix L. Paulauskas, Fred S. Baker, C. Cliff Eberle, and Amit Naskar. 2008. "Multi-task Research Program to Develop Commodity Grade, Lower Cost Carbon Fiber." *40th International SAMPE Technical Conference - Memphis, TN - Sep 8 - 11, 2008.*

Xu, Lanhong, Tom Mase, and Lawrence T. Drzal. 2003. "Improving Adhesion between Carbon Fibers and Vinyl Ester Resins." SPE ACCE, Troy, MI (http://www.speautomotive.com/aca).

Chapter Three

Carbon Fiber Composite Construction

I am an enthusiast, but not a crank in the sense that I have some pet theories as to the proper construction of a flying machine. I wish to avail myself of all that is already known and then, if possible, add my mite to help on the future worker who will attain final success.
—Wilbur Wright

3.1 The Splendid Variety of CFCs for Automotive Applications

Carbon fiber composite (CFC) is not a material — it is a material *family* that is composed of a variety of carbon fiber types, fiber reinforcement constructions, and resin (matrix) systems. The fibers provide reinforcement to the matrix, much like steel reinforcing bar or mesh reinforces concrete. The extremely high tensile stiffness and strength of carbon fiber result in a composite that has high stiffness and strength. The resin, or matrix material, binds the fibers together to contribute to other mechanical properties such as toughness, as well as physical characteristics such as resistance to environmental conditions, including fire, ultraviolet (UV) light, and chemical or moisture attack.

Carbon fiber reinforcements have traditionally been continuous fibers produced from a polyacrylonitrile (PAN) precursor, as adopted from the aerospace industry. Because of the high cost of manufacturing PAN-based CF, there have been considerable efforts put toward non-PAN based precursors, including pitch-based, polyolefin-based, and bio-fiber based. Pitch-based CF is commercially available and provides some unique properties; however, the pitch-based CFs on the market today provide no significant cost reduction compared to PAN-based CF.

Regardless of the precursor source, practical use of carbon fibers requires they be used as a bundle as opposed to individual fibers. Individual continuous carbon fiber filaments are collimated, with or without a twist (but usually without a twist) into a strand, which is called a tow in the carbon fiber industry. As explained in Chapter 2, the number of filaments in the tow is called the tow size, and is designated in terms of Ks, where 1 K is one thousand filaments. Commonly produced tow sizes range from 1K to 24K, with filament diameters no greater than 0.010 inch (2.54×10^{-4} m) and usually much less (e.g., 7×10^{-6} m).

Carbon fiber in continuous form provides the most benefit in mechanical properties. Continuous fiber can be used directly in a filament winding process, in which controlled tension is applied to the filaments as they are wound in a geometric pattern to form a structure. Continuous fiber can also be processed into a unidirectional tape, in which the fibers (in tow form) are oriented in one direction to make a continuous tape of a set width, or they can be processed into a weave, in which the filaments are interlaced into a fabric pattern that is commonly a 0°/90° grid but can be made on the bias at +45°/-45°. Other forms of continuous carbon fiber include three-dimensional weaves, knits, stitched fabrics, or mats of loosely gathered (nonwoven) rovings.

Discontinuous or chopped fibers can provide a significant portion of the tensile modulus and strength benefits that continuous carbon fiber offers, depending on the length of the discontinuous fibers and processing method. Discontinuous fibers are typically used for less demanding structural applications. Discontinuous fibers exist in parts made using a sprayup process, in which continuous carbon fiber tow (strand) rovings are chopped to a specified length in a chopper gun and sprayed into an open mold along with resin and catalyst. Discontinuous or chopped fibers can also be used in resin transfer processes, injection molding, and direct long fiber thermoplastics.

The matrix material or resin has traditionally been epoxy adopted from the aerospace industry. However, three main drivers continue to push resin development for the automotive industry: cycle time, performance (mechanical and physical properties), and processing methods. Traditional epoxies have been modified for performance, for example, higher toughness and/or faster cure time. Common aerospace-grade epoxies require 30 minutes or longer to cure, which can only be accommodated in small production volume, i.e., niche vehicles. Mass production of mainstream vehicles requires cure times in the single digits. Aside from modifying epoxies to reach these faster cure times, other thermosetting resins have been developed to meet this need, including polyurethane (PU) systems and polydicyclopentadiene (pDCPD). Thermoplastic resins, which can be used in certain processes for manufacturing CFCs, do not require a curing stage and hence can provide rapid cycle times that are attractive to the automotive industry. Thermoplastic resins that can be used in CFCs range from the commodity-type resins such as polypropylene and nylon, to higher-end resins such as polyphenylene sulphide and cyclic polybutylene terephthalate (cPBT), depending on the required material performance and processing method.

The amount of fiber reinforcement and how the fibers are integrated in the matrix strongly affect the mechanical properties of the composite. The ratio of fiber-to-resin, or percentage fiber to volume, determines the density of the composite and hence the final weight of the part. The fiber volume, in conjunction with the fiber form and composite construction, determine the mechanical properties of the composite. Fiber volumes can be as low as 20% for low-cost, nonstructural components, and as high as 70% in some high-end structural components. A fiber volume of 35%–60% is common in automotive carbon fiber components. The composite construction, or how the fibers are oriented and architecturally arranged, should be designed according to the directions and modes of loading on the final part. If there is a dominant primary load on the part, the fibers should be dominantly aligned in the same direction to realize the stiffness and strength attributes of carbon fiber. In panel-type geometries, common fiber orientations are 0°, 90°, +45°, and -45°. For beam/column and vessel-type geometries, fiber orientations are usually ±33° to ±45°. A 54° winding angle satisfies both the circumferential (hoop) and longitudinal (axial) strength requirements of most pipes and pressure vessels, usually manufactured by the filament winding process. However, if more stress is placed on the pipe in the axial direction, as is the case with an unsupported span, a ±20°/±70° fiber orientation will provide a stiffer bending modulus for increased axial strength [Part 2007].

The intended fabrication method also will influence design. For example, manufacturers of filament-wound or tape-laid structures use different reinforcement forms and buildup patterns than those used either for laminate panels laid up by hand or for vacuum-bag-cured prepreg parts. Resin transfer molding (RTM) accommodates three-dimensional preforms more easily than do some other manufacturing techniques [Fabrication 2007].

3.2 Fiber Reinforcement Forms

3.2.1 Bonded

In bonded reinforcements, the fibers are held together by a polymer, ranging from a small amount of binder to semi pre-impregnated, to fully pre-impregnated (prepreg) mats. The fibers can be chopped, such as in chopped strand mats, or continuous, such as in continuous filament random swirl mats. Bonded reinforcements can also be in the form of unidirectional fibers in a thermoplastic prepreg. Bonded mats containing a small amount of binder holding the fibers in place are processed into reinforced composites using a liquid composite molding process. Mats in the form of thermoplastic prepregs can be processed by thermoforming or compression molding depending on the level of pre-impregnation.

3.2.2 Unidirectional Tapes

In unidirectional tapes, continuous fibers of a set width are all aligned in one direction. The fibers are impregnated with resin and backed with a polymeric or paper backing. Unidirectional tapes are typically used in a laminated construction, with layers of tape stacked up to create the desired thickness. Because carbon fiber has much higher tensile modulus and strength in the axial direction than in the transverse direction, the orientation of the tape layers can be, and usually is, varied through the stack to provide the required mechanical properties in the end composite.

An investigation by Feraboli, Masini, and Friedman [2002] looked at the effect of fiber architecture on the strength of carbon fiber/epoxy composite panels. Architectures studied were unidirectional, multidirectional, quasi-isotropic and cross-ply, 2×2 twill, and eight-harness satin weave, and performance was evaluated in flexure and inter-laminar shear. The unidirectional ranked far above the other architectures in flexural strength and essentially equivalent to the cross-ply and quasi-isotropic. The two woven architectures ranked lowest in flexural strength and inter-laminar shear strength. However, other considerations may not favor unidirectional tapes: unidirectional tape laminates are more prone to the formation of voids during consolidation because of their tight fiber arrangement; tapes typically are more susceptible to external agents such as moisture and localized impact; and tapes are more prone to thermal delamination [Feraboli, Masini, and Friedman 2002].

3.2.3 Stitched

Stitched fabrics are formed by sewing or knitting a lightweight fiber as a loop around the reinforcement tow. Because the lightweight fiber does not crimp the tows, allowing them to stay in-plane and aligned, stitched fabrics are also referred to as non-crimp fabrics. Stitched fabrics can be made up of a single layer or multiple layers, and multiple layers may have different orientations of the reinforcing tows, similar to how unidirectional tapes can be laid up at different orientations (0/90, ± 45, 0/45/-45/90, etc.). Stitched fabrics can have a higher fiber orientation factor than woven and can also pack to provide high fiber volume fractions. On the flip side, this can hinder resin penetration. An image of stitched fabric is shown in Fig. 3.1 [Composites 2011].

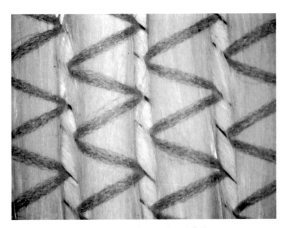

Figure 3.1. Example of stitched fabric.
(Image courtesy of John Summerscales, Plymouth University, UK)

3.2.4 Knits

In warp knits, each stitch within a row has a separate thread. In weft knits, there is one yarn per row. Weft knits provide more stretch, but warp knits are harder to unravel.

Warp knitting is a very versatile fabric production, being able to produce flat, tubular, or even three-dimensional fabrics that can have either an open or closed structure.

In weft- or warp- insertion knits, a fabric tow follows a meandering path of symmetric loops to produce a fabric that is more easily stretched than a weave. The horizontal row of loops is called the *course*, and the vertical lines are called the *wakes*. The principle of weft-insertion warp-knitting is to provide reinforcement fibers in parallel across the whole width of the fabric.

Takahashi et al. report on the development of a carbon fiber knitted fabric and unique resin with good de-aeration performance [Takahashi, Kageyama, and Kawamura 2011]. Their benchmark was woven fabric used in F1 that is formed by weaving the warp and weft threads so the fiber bundles do not become separated. Their newly developed material uses a multi-axial knitted fabric in which the fiber bundles are laid out in straight lines in multiple layers and stitched together to prevent separation. Large-diameter fiber bundles with high areal weight were used to reduce the number of layers and take advantage of the directional stability of the multiaxially oriented fibers. They also developed a prepreg structure that provided pathways in the knitted fabric for air to escape under low pressure, as well as semipreg material with a modified epoxy resin with optimized flowability [Takahashi, Kageyama, and Kawamura 2011]. The material is currently used in Toyota's flagship sports car, the Lexus LFA.

3.2.5 Wovens

Woven fabric is produced on a loom by interlacing two sets of orthogonal fiber bundles (tows). The interlacing can occur at each orthogonal fiber tow, or at tow intervals which may or may not be the same in both directions. Different tow sizes may also be used within one direction and may be different in the two orthogonal directions. The various interlacing patterns produce the different weave patterns, with the simplest being the plain weave. Other common patterns are twill weave and five-harness satin weave. A fabric with equal number and weight of tows in the two orthogonal directions is termed *balanced*. For some applications, particularly structural, the majority of the tows may be oriented in one direction with a small quantity of lighter orthogonal fibers to produce a unidirectional fabric.

The waviness of the fibers in the weave is called the *crimp*. The simplest way to express the crimp is by the number of crimps or waves in a unit length. Crimp can also be expressed as the difference in distance between two points on the fiber bundle in the crimped state and the distance between the same two points when the fiber bundle is straightened under suitable tension. As the crimp increases, drape and in-plane permeability increase, but mechanical properties of the end composite tend to decrease.

The vast majority of woven carbon fibers are two-dimensional planar fabrics. However, it is possible to produce triaxial or three-dimensional weaves. Because they are three-dimensional by nature, there is no interlaminar weakness in 3-D woven constructions. Three-dimensional fabrics such as 3-D weaves (or 3-D braids) are primarily used as preforms and consolidated in a liquid composite molding process (Chapter 4).

Using automated continuous textile performing machinery, a unit of thick multi-layered fabric is formed during each weaving cycle. The process developed by 3TEX Inc. to create 3-D woven composites, called 3WEAVE™, has automated multiple weft insertion in a single weaving cycle, automated production of thick and/or net-shaped forms having various cross-sections or core or pile structures, and the ability to include up to one-third of the fiber weight as Z-direction (through thickness) fiber in controlled amounts [Dickinson 2002]. Figure 3.2 shows schematics of thin and thick 3WEAVE™.

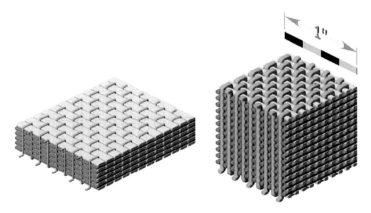

Figure 3.2. Schematics of thin and thick 3WEAVE™ 3-D orthogonal woven preforms [Dickinson 2002].
(Image used with permission from SAMPE)

Figure 3.3. Examples of net-shape and thick one-piece 3WEAVE™ preforms of carbon, aramid, and E-glass [Dickinson 2002].
(Image used with permission from SAMPE)

The woven sections are consolidated using vacuum-assisted resin transfer molding (VARTM) (discussed in Chapter 4). Examples of net-shape thick 3WEAVE™ preforms of carbon, aramid, and E-glass fibers are shown in Fig. 3.3. The non-crimp combined with Z-direction fibers permit these materials to be machined with much less damage to the remaining component than is possible with laminates. Several interesting examples of machined parts are shown by Dickinson [2002].

Combining 3-D-weaving with structural adhesive bonding led to a JEC 2010 Process Innovation Award for Biteam and its partners in the Modular Joints for composite aircraft components (MOJO) project [Biteam 2010]. Their innovations combined non-crimp carbon fabrics and tailored preform profiles — including out-of-plane reinforcements — consolidated in out-of-autoclave infusion, with structural bonding processes to fabricate modular joints that have damage tolerance. Compared to riveting, structural bonding can provide about 25% overall cost saving, up to 60% assembly cost saving, and about 50% weight savings [Biteam 2010].

3.2.6 Braids

Braiding is the process of interlacing three or more yarns in such a way that they cross one another but no two yarns are twisted around one another. The yarns are laid in a diagonal formation (continually woven on the bias) to create flat, tubular, or solid constructions. Braiding can be done two-dimensionally to produce a fabric or three-dimensionally to produce a construction. Braided fabric can be dry or prepregged. Prepregged braids can be consolidated using a vacuum bag process or two-piece mold.

Conventional uses of 2-D tubular braids have typically required closed-mesh braiding. The reason for that is due in part to the difficulty to produce open-mesh braids, and in part to the lack of adequate tools to model the large deformation behavior of open-mesh braids. In work by Carey, Ayranci, and Romanyk [2010], they showed that the longitudinal elastic and shear moduli were overpredicted by current models, and that open-braided structures have a larger window of tailorability and hence potential applications in stiffness-critical designs than had been previously thought.

The principal of manufacturing complex net-shaped preforms based on 2-D braiding is discussed by Brandt, Drechsler, and Gessler [2001]. The example presented is a J-stiffener with a curved profile. Its manufacture entailed overbraiding on a curved mandrel, allowing longitudinal yarns to be laid up straight and to correct length, which are different in foot and flange of the J due to the curved profile. The mandrel is removed, and the braided preform is folded into the correct J profile preform, ready for impregnation.

3.3 Constructions
3.3.1 Laminates
Laminates are constructed from layers of single-ply material. The single plies are typically unidirectional tapes, weaves, or stitch-bonded fabrics, making the majority of laminates essentially two-dimensional materials with excellent mechanical properties in the plane of each lamina (ply) but weak between the layers.

An understanding of layered or laminated structural behavior is vital to effective composite component design. Adhesion between laminate layers (called plies) is critical; poor adhesion can result in delamination under stress, strain, impact, and load conditions. Ply layup designers must consider mechanical stresses/loads, adhesion, weight, stiffness, operating temperature, and toughness requirements, as well as variables such as electromagnetic transparency and radiation resistance. Additionally, composite component design must encompass surface finish, fatigue life, overall part configuration, and scrap or rework potential, to name just a few of the many applicable factors.

The layup schedule can affect the microcracking performance of a commercial prepreg laminate system when exposed to thermal cycling. Wide variations in temperature can occur in aerospace components during takeoff and landing, or in automotive under-hood applications. These thermal cycles can cause microcracks in the matrix parallel to the fibers due to mismatches in fiber and matrix properties and anisotropic thermal expansion coefficients, as revealed in a study by Dharia, Hayes, and Seferis [2001]. Microcracks result in reduced mechanical properties and can lead to delamination, fiber breakage, and uptake of moisture or other fluids. In that study, commercially available carbon fiber prepreg laminates were produced with different layup schedules, made to different thicknesses, and subjected to different thermal ramp rates. It was found that the sensitivity of a given composite system to thermal microcracking can be adjusted by the layup schedule as well as the laminate thickness.

Sandwich construction, a common type of composite structure, combines a lightweight core material with laminated composite skins (facesheets), similar to the construction of corrugated cardboard. These very lightweight panels have the highest stiffness-to-weight and strength-to-weight performance of all composite structures and are extremely resistant to bending and buckling. Suitable core materials include closed-cell foams, balsa wood, and celled honeycomb in a variety of forms (aluminum, paper, or plastic). Some foam cores are syntactic, containing hollow microspheres to reduce weight. Sandwich construction is used extensively on modern aircraft and boats as well as in applications such as cargo containers and modular buildings. Some material suppliers offer reinforcements with an integral core, such as a core-like material (e.g., foam rods) stitched together with glass or carbon fibers. A woven or unidirectional fiber form integrated with a core material provides a unitized composite structure that is amenable to both infusion and closed-mold processing.

3.3.2 Filament Winding

Filament winding is a process in which fiber tows are wound around a mandrel or core over a mandrel. Filament winding can be done with dry tows in a wet winding process, or with prepreg tows, but the delivery systems differ. Major differences include the need for higher fiber tension at both the spool and the part in prepreg winding compared to wet winding. However, the carbon dust particle and solvent fume extraction systems needed in wet winding are not required in prepreg winding [Ericksen 2011]. The need to constantly control the fiber tension over the complete path of the fiber is key to producing repeatable quality products in either process. Figure 3.4 shows a power winding head fabricating a vessel.

Figure 3.4. Filament winding delivery with power winding head.
(Image courtesy of TCR Composites)

3.3.3 Stitching Preforms

Stitching in combination with fiber preforms can allow different textiles (braids, wovens) to form an integrated preform. It can also improve the mechanical performance by increasing the delamination resistance of a laminate [Brandt, Drechsler, and Gessler 2001].

Conventional lock and chain stitches are limited to flat preforms because they require access to both sides of the material. One-sided-stitching devices that manipulate robotically have been developed that permit stitching to assemble complex 3-D-shaped preforms. Stitching displaces the fibers, causing a reduction of in-plane properties, the degree of which depends on the stitching technique and parameters [Brandt, Drechsler, and Gessler 2001]. Figure 3.5 shows a one-sided stitching robot at the European Aeronautics Defence and Space Company (EADS).

3.3.4 Three-Dimensional Braids

Three-dimensional braiding provides great flexibility in producing 3-D fiber architectures and continuous performs. Several companies provide braiding machines, which are fully automated and computer-controlled. Tubular braiding is well established for producing three-dimensionally shaped performs [Brandt, Drechsler, and Gessler 2001]. It is also possible to manufacture complex three-dimensional preforms based on 2-D braiding, by braiding yarns on cores, removing the core, and then folding the braiding to make, for example, stinger shaped preforms, as shown in Fig. 3.6.

Some machines are capable of changing the shape of the preform on the fly by transitioning through different braiding patterns. This permits shapes such as T-sections

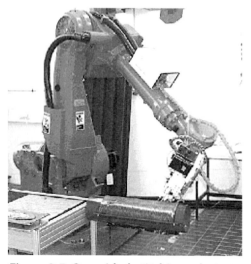

Figure 3.5. One-sided-stitching robot at EADS
[Brandt, Drechsler, and Gessler 2001].

Figure 3.6. Curved J-stiffener manufactured from preform based on 2-D braiding
[Brandt, Drechsler, and Gessler 2001].

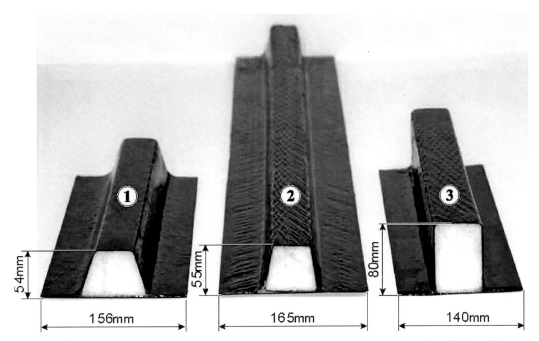

Figure 3.7. Integral composite Hat-beams fabricated with the same 3-D braided preform
[Mungalov, Duke, and Bogdanovich 2006]. (Image used with permission from SAMPE)

to be made (see Fig. 3.7), creating a joint-free transition where laminates and other forms of constructions would consist of two separate components joined together. 3TEX's patented 3-D braiding named 3BRAID® consists of a three-dimensional rotary machine with a nine-module braider that can handle up to 576 braided yarns and 144 axial tows to produce relatively large components. Probably the most familiar and exciting use of 3-D braiding technology to date is in the Lexus LFA [Middlehurst 2011].

3.3.5 Nanostitching and Fuzzy Fibers

Nanotechnology developments are underway in many industries and applications, including CFCs. One of the ongoing developments is the use of nano-engineered composites, using aligned carbon nanotubes (CNTs) to provide tailored reinforcement of existing CFCs. Two different approaches being developed by Wardle et al. [2010] are nanostitching and fuzzy fibers. In nanostitching, CNTs between lamina are aligned perpendicularly to the lamina to reinforce the weak interlaminar interfaces that exist in composite laminates. In their "fuzzy fiber" developments, they create nanofibers that extend radially from the carbon fiber surface (perpendicular to the carbon fiber surface), making the carbon fiber appear "fuzzy." The fuzzy fibers can then be woven into a fabric. The processing of such a nano-engineered carbon fiber composite is enabled by capillarity-induced wetting of these aligned nanofibers. Figure 3.8 shows schematics of both nanostitching and fuzzy fibers.

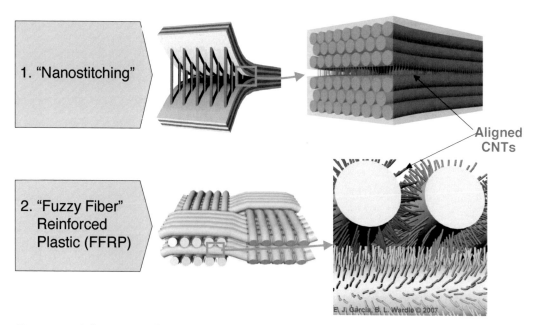

Figure 3.8. Schematics of nanostitching and fuzzy fibers [Wardle et al. 2010].
(Courtesy of E. J. Garcia and B. L. Wardle)

3.4 Closing Thoughts

Selecting the right fiber, resin, and construction for a particular application requires knowledge of not only the fiber and resin material properties, but also of the method of manufacturing. The method of manufacturing influences the composite construction and end properties, while the surface quality of the finished part (Class A or non-Class A) and the production volumes to be made in turn dictate what manufacturing methods are technically and economically viable. One must also factor in the commercial competitiveness with other materials with regard to vehicle installation, maintenance, and lifecycle issues. There is usually more than one way to make a carbon fiber composite automotive part, and all factors should be considered to make the best decision.

The development of automation in the manufacture of composites, the processing of their precursors, and the shortening of production cycles that achieve consistent quality in serially produced parts are real challenges in the CFC industry, which all players in the industry recognize.

The Composites Automation Development Center opened by Web Industries [Web Industries 2011] in the Atlanta, Georgia, area will help customers test new CFCs for use in automated fiber placement processes. This is the first testing and development center dedicated to helping raw material suppliers, fabricators, and manufacturers leverage these new materials in automated manufacturing.

Similar centers exist in Europe. The Composite Structures Development Centre located at the Airbus facility in the UK is concentrating on out-of-autoclave manufacture of large CFC structures, while GKN's Low-Cost Composites Manufacturing Centre is working on automated production methods to lower the cost of CFC structures by 30% [Balzer et al. 2008].

There is a tremendous amount of science, engineering, and technology behind carbon fiber composites. The interested reader thirsting for detailed technical knowledge is referred to some select references: *Carbon fibers and their composites*, Peter Morgan, CRC Press, 2005; *Carbon fiber composites*, Deborah D. L. Chung, Butterworth-Heinmann, 1994; *Fibers and composites*, Pierre Delhaès, CRC Press 2003; *Design and control of structure of advanced carbon materials for enhanced performance*, B. Rand, S.P. Appleyard, M. F. Yardim, Springer 2001; *Carbon materials science and engineering – From Fundamentals to Applications*, Michio Inagaki, Feiyu Kang, Tsinghua University Press, 2006; and *Composites for automotive, truck and mass transit: materials, design, manufacturing*, Uday Vaidya, DEStech Publications, Inc. (Google eBook), 2011.

References

Balzer, Brian B., Jeff McNabb, David Stienstra, and Thomas Mensah. 2008. "Impact of a Faster Curing Process on Selected Properties of Carbon Fiber Prepreg Matrix Using an Industrial Microwave Process." SAMPE 2008 - Long Beach, CA May 18 - 22, 2008.

"Biteam Shares JEC 2010 Process Innovation Award." 2010. *netcomposites* (http://www.netcomposites.com/newspic.asp?5979 Accessed August 28, 2011).

Brandt, Jürgen, Klaus Drechsler, and Andreas Gessler. 2001. "Comparison of Various Braiding Technologies for Composite Materials in Aerospace Applications." SAE Paper No. 2001-01-2626. SAE International, Warrendale, PA. 2001.

Carey, Jason, Cagri Ayranci, and Daniel Romanyk. 2010. "Tailoring 2D Large-Open-Mesh Braided Composites." Society of Plastics Engineers, *Plastics Research Online* no. 10.1002/spepro.002964.

Composites Design and Manufacture (BEng) - MATS 324 Reinforcement Fabrics. 2011. ACMC, School of Marine Science and Engineering (http://www.tech.plym.ac.uk/sme/MATS324/MATS324C2%20fabrics.htm Accessed 2011).

Dharia, Anand K., Brian S. Hayes, and James C. Seferis. 2001. "Evaluation of Microcracking in Aerospace Composites Exposed to Thermal Cycling: Effect of Composite Lay-Up, Laminate Thickness and Thermal Ramp Rate." 33rd International SAMPE Technical Conference - Seattle, WA - November 5 - 8, 2001.

Dickinson, Larry C. 2002. "Thick 3D Woven Composites as a Standard Material: Manufacturing, Properties and Applications." SAMPE 2002 - Long Beach, CA May 12 - 16, 2002.

Ericksen, Leo, D. 2007. "Prepreg Winding Delivery Systems." TCR Composites (http://www.tcrcomposites.com/pdfs/winding_delivery_ppt.pdf Accessed August 27, 2011).

"Fabrication methods." 2007. *Composites World* no. November 1, 2007. (http://www.compositesworld.com/articles/fabrication-methods Accessed August 1, 2011).

Feraboli, Paolo, Attilio Masini, and Keith Friedman. 2002. "Considerations on 6 Fiber Architectures of a Carbon/Epoxy Composite in the Design of a Vehicle Body." SAE Paper No. 2002-01-2037. SAE International, Warrendale, PA. 2002.

Middlehurst, Tony. "Wonder Material." (https://secure.drivers.lexus.com/lexusdrivers/magazine/articles/Vehicle-Insider/Carbon-Fiber Accessed August 27, 2011).

Mungalov, D., P. Duke, and A. Bogdanovich. 2006. "Advancements in Design and Manufacture of 3-D Braided Preforms for Complex Composite Structures." 38th International SAMPE Technical Conference, November 6-9, 2006.

"Part design criteria." November 1, 2007. *Composites World* (http://www.compositesworld.com/columns/part-design-criteria).

Takahashi, Naoyuki, Yuji Kageyama, and Nobuya Kawamura. 2011. "Research of Multi-Axial Carbon Fiber Prepreg Material for Vehicle Body." SAE Paper No. 2011-01-0216. doi: 10.4271/2011-01-0216. SAE International, Warrendale, PA. 2011.

Wardle, Brian L., et al. 2010. "Bulk Nanostructured Materials and Advanced Composites." Nano-Engineered Composite Aerospace Structures Consortium (2010 Gordon Research Conference on Composites, Ventura, CA, January 20, 2010).

"Web Industries Opens Composites Automation Development Center for Testing CF Composites." 2011. *Omnexus SpecialChem, August 9, 2011.* (http://www.omnexus.com/news/news.aspx?id=28433 Accessed August 9, 2011).

Chapter Four

Manufacturing Processes for Carbon Fiber Composites

If you need a new process and don't install it, you pay for it without getting it.
—Ken Stork

4.1 Introduction

The manufacturing of CFCs starts with the constituent raw materials being processed into a fiber construction (with or without resin), which is then processed into a component via consolidation. Figure 4.1 schematically shows this progression for a simplified array of a few materials and processes.

The traditional manufacturing method to produce carbon fiber composites for automotive applications is the autoclave, adopted from the aerospace industry. The autoclave is a pressurized oven that subjects the uncured fiber/resin construction to the pressure that is needed for consolidation and the temperature that is needed for curing and hardening the resin.

Various other methods have been developed over the decades and are in commercial use today, primarily for non-aerospace applications including automotive, although the aerospace industry has also been pursuing out-of-autoclave methods in recent years.

The most widely used non-autoclave methods include pultrusion, bulk molding compound (BMC), sheet molding compound (SMC), compression molding, injection molding, direct in-line compounding, resin transfer molding (RTM) and its variants, filament winding, automated tape layup/automated fiber placement, spray forming with infusion, forging, and hot oil cure (e.g., Quickstep® by Quickstep Holdings Limited and Pressure Press (provisional patent issued) by Plasan Carbon Composites). Although there is some application overlap across these methods, the fiber form, resin type, composite construction, and desired material properties directionally guide the manufacturing method selected.

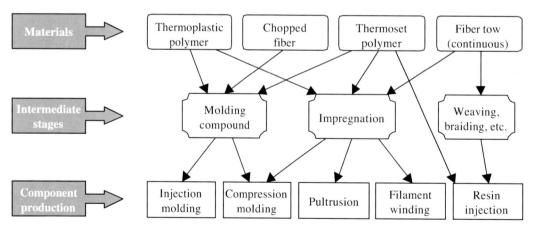

Figure 4.1. Comparative Manufacturing Processes for CFCs [Industry 2000].
(Courtesy of EPRI)

4.2 Injection Molding

Injection molding has been used for decades to manufacture fiber reinforced plastics. In the automotive industry, injection molding is used with both thermoplastic and thermoset resins, with thermoplastic resins having dominated the industry due to their lower cost to achieve the required material properties. Although glass fibers and natural fibers are by far the most common fiber reinforcements used in injection molding in the automotive industry today, the process is equally suitable for carbon fibers.

Injection molding of non-reinforced plastic is a relatively simple process in which pellets of solid resin are fed through a hopper into an injection screw where they become melted under controlled temperature, pressure, and time conditions, and the melt is subsequently injected into a mold cavity where it cools and solidifies. Injection molding of fiber reinforced plastic is done in a similar manner using pellets that contain the fiber reinforcement within the resin. These fiber reinforced pellets typically contain 10–35% fiber by weight, and are manufactured in a separate extrusion or pultrusion process by feeding discrete (short) lengths of fibers into melted resin, then extruding or pulling the reinforced resin through a die and pelletizing the material. The fiber lengths in pellets historically were approximately 5 mm or less. Because longer reinforcing fibers provide higher mechanical properties, the processing technology evolved to achieve longer fibers in the fiber reinforced pellets. As a result, there is a distinction made between "short fiber" and "long fiber" reinforced plastics in the automotive industry; definitions vary, but pelletized long fibers typically have fiber lengths around 10–25 mm. The fiber lengths in the finished injection molded part are typically less than 1 mm and often less than 0.5 mm for short fiber reinforced plastics, and typically less than 5 mm for long fiber reinforced plastics.

In a general Long Fiber Thermoplastic (LFT) process, resin from an in-line compounder is fed into a mixing extruder as fibers are also fed in, and the mix is either injection molded or produces a LFT plaque that will be consolidated through compression molding.

With the desire to further improve the mechanical properties of injection molded fiber reinforced plastics, technology advancements have been made over the past decade to create longer fiber lengths in the finished part. The technology developed to achieve in-line compounding is often referred to as Direct Long Fiber Reinforced Thermoplastic or D-LFT. Other acronyms are Endless D-LFT (E-LFT) for direct extrusion with continuous fiber reinforcements from in-line compounding, and Structural D-LFT (S-LFT) for injection molding from in-line compounding. The D-LFT for compression molding provides longer fiber lengths, from 10 mm up to 80 mm (approximately 0.5 to 3 in), but has limited three-dimensional capabilities and associated post-mold trimming requirements. A new variant of injection-LFT presented by du Toit et al. [2010] uses a closed molding process similar to LFT injection but offers more 3-D capability and longer post-mold fiber lengths. Lengths of 10 mm (0.4 in) are typical in complex shapes, while lengths up to 50 mm (2 in) can be attained in simple structures. This produces parts with mechanical properties closer to D-LFT with compression molding.

Although injection molding has been traditionally focused on thermoplastics, BMC injection molding has become popular because of its dimensional stability. In automated injection molding of BMC, a ram forces a metered shot of material through a heated barrel and injects it under high pressure (5000–12000 psi) into a closed, heated mold [Fabrication 2007]. BMC injection molded parts require minimal finishing, and injection speeds are typically 1–5 seconds. BMC can also be compression molded or transfer molded to produce parts with thick sections.

4.3 Compression Molding

Compression molding, as with injection molding, is a high-volume method. Metal dies are used to handle the pressures and high-volume part production. Compression molding is typically used for thermoset glass SMC, but the automotive industry has been exploring the use of carbon fiber SMC [Fabrication 2007]. Additionally, Teijin's carbon fiber reinforced thermoplastic compression molding technology received the Frost & Sullivan Global Automotive Carbon Composites Technology Innovation Award in 2011 [Teijin 2011].

Feraboli, Graves, and Stickler [2007] studied the mechanical performance of compression molded randomly distributed carbon fiber prepreg chips. Stiffness performance was almost on par with the continuous fiber benchmark, but ultimate strength was considerably reduced.

Continuous compression molding (CCM), developed in the early 1990s, is an automated, semi-continuous process that can produce flat panels of almost unlimited length or highly shaped profiles, with void content routinely less than 1% [Gardiner 2010]. Thermoplastic prepreg tape is laid, heated, compressed, and consolidated into a sheet. Upon consolidation the sheet is cut to size and put into a two-sided hydraulic stamp-forming press where it is heated and stamped to shape.

4.4 Thermoforming

Thermoplastic preforms can be fabricated into a part by a stamping-like process called thermoforming. Thermoforming can create complex geometries, but to minimize waste the material properties and temperatures need to be predicted accurately prior to forming. An investigation by Smith and Vaidya [2010] determined that variations in heat fluxes through the thickness are bound to exist, but are minimized with constituents having high thermal conductivity.

In Fibreforge's process, fabrication of thermoplastic composite parts centers around a flat net-shape preform comprised of layers of continuous fiber reinforced thermoplastic, which they refer to as a "tailored blank" [Burkhart and Cramer 2006]. In a fast and easy process, the tailored blanks are consolidated and formed into their final shape through a single thermoforming step.

4.5 Sheet and Strand Molding Compound

SMC is made by dispersing long strands of discontinuous fiber on a bath of resin. Resin paste is dispensed onto a plastic carrier film, and fiber rovings are cut and dispersed on top. Once the fibers have settled through the depth of resin paste, another film is placed on top to create a sheet that sandwiches the fibers. The sheet is rolled and stored at a specific temperature while it matures. When ready for use, the carrier film is removed and the material is cut into charges and placed in a die. Heat and pressure, such as provided through compression molding, consolidate and cure the composite. SMC can also be processed using a direct molding technology, which essentially forms a sheet that is immediately processed (cut, formed, and compressed) into a finished component all in the same production line [Henning 2010].

Direct Strand Molding Compound (D-SMC) [Potyra et al. 2009] provides a continuous flow of raw material to a finished compression molded thermoset part. Resin and continuous fibers (dosed with resin) are processed through a twin screw extruder to form charges that have been cut to size to be placed in a compression mold to create a part.

4.6 Spray Forming

Spray forming was traditionally done with chopped fibers. Resin is pumped to a robotic dispensing unit, and continuous reinforcing fibers are fed through a cutter that chops them to the desired length. The chopped fibers and resin combine in the head and are sprayed simultaneously into an open cavity of the mold. Alternatively, chopped fibers can be sprayed with a powder binder onto a perforated tool (under vacuum). Developments in recent years have enabled continuous fibers to be sprayed [Henning 2010].

Dodworth [2009] presented a novel spray depositing technique for compression molding structural components from discontinuous fibers for medium volume applications. The novelty entails using magnetic fibers with the carbon fibers to hold the fibers on the tool surface to permit high precision in fiber placement.

4.7 Pultrusion

In the pultrusion process, fiber reinforcements are pulled through a guide plate that helps locate the reinforcements correctly in the final part. The aligned reinforcements then pass through a resin impregnation chamber that contains the polymer solution, consisting of the resin, catalysts, fillers, and any other additives. Most pultrusion systems use electric or hot oil systems to heat separate heating zones on the die surface.

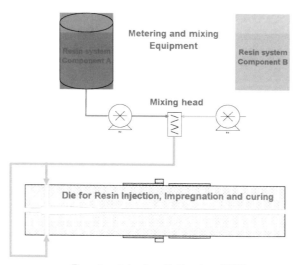

Reaction Injection Pultrusion (RIP)

Figure 4.2. Schematic of a pultrusion line [Henning 2010].
(First published in the 2010 SPE ACCE, Troy, MI)

The number of heating zones depends on the type of heat source, the length of the die, the resin type, and the speed of the process. Radio-frequency heating units placed after the resin impregnation step, and before the curing die, can be used for longitudinally glass reinforced pultrusions, but cannot be used with carbon fiber reinforcements because of potential fire hazards [Industry 2000]. Pultrusion is very well suited for structural components subjected to a predominantly axial or bending load. A schematic of a reaction injection pultrusion line is shown in Fig. 4.2.

4.8 Filament Winding

In filament winding, a fiber application machine places fiber on a mandrel in a predetermined configuration. Computer-controlled filament-winding machines are available that have 2 to 12 axes of motion [Fabrication 2007]. For thermoset systems, dry fiber can be passed through a resin bath just prior to placement on the mandrel, or pre-impregnated fiber can be used to eliminate the resin bath. Alternatively, in dry winding, fibers are wound to create the dry shape, which is then used as a preform for a liquid infusion process like RTM. The mandrel can remain part of the finished composite or it can be removed after curing. To make complex parts, mandrels may be made to be collapsible, disassembled, or dissolved [Fabrication 2007]. For thermoplastic systems, the fibers start prepregged and are heated as they are wound on the mandrel. The prepreg is heated, placed, compacted, consolidated, and cooled in a single operation [Fabrication 2007]. Because of the resulting fiber architecture, filament-wound components have exceptional hoop strength.

Acrolab developed a heated mandrel to cure tubes while still on the filament winding machine. A heatpipe thermally enhanced (HPTE) mandrel is heated by an induction power supply and induction coil, enabling curing with only thermal energy provided by the mandrel [Ouellette 2010].

Automated fiber placement (AFP) and automated tape placement (ATP) consist of automated placement of prepreg tows (AFP) or prepreg tapes (ATP) around mandrels or into tools to increase processing speed and reduce scrap. ATP is typically faster than AFP, but AFP is better for contoured surfaces [Fabrication 2007]. Both processes are used extensively in aerospace. A nice description of automated tape laying methods is presented by Grant [2005].

4.9 Resin Infusion Processes

In resin transfer molding (RTM), a dry fiber reinforcement preform is placed in position and enclosed in the mold. Liquid resin is injected into the mold under relatively low pressure. To provide good heat transfer, metal molds are usually used, but composite molds can be used. Figure 4.3 shows a schematic of an RTM apparatus. Extremely low viscosity resin is needed for thick parts so it can quickly and evenly permeate through the preform before curing. Traditionally, RTM has used two-part epoxy that is mixed just prior to injection. In contrast, in reaction injection molding (RIM), a rapid cure resin and a catalyst are injected separately into the mold, and the mixing and chemical reactions occur in the mold instead of the dispensing head. Structural RIM (SRIM) is used to produce structural parts that do not require Class A finish.

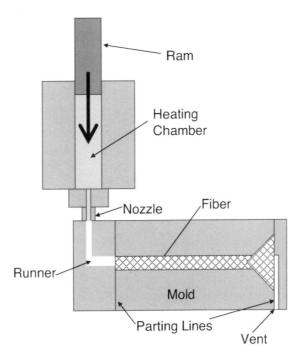

Figure 4.3. Schematic of an RTM apparatus
[Buckley et al. 2006].

There are several variants of RTM, including vacuum-assisted RTM (VARTM), RTM-Light, Compression RTM, and Core Expansion RTM. RTM-Light uses low injection pressure with vacuum to permit the molds to be lighter weight and less expensive. VARTM is accomplished at low pressures and frequently at low temperatures, which often enables low-cost tooling to be used [Norris et al. 2000]. In VARTM, the resin is drawn into the preform by vacuum as opposed to pressure. The preform is placed in a one-sided tool and a cover is placed over top to create a vacuum-tight seal. Resin is drawn in by vacuum through strategically placed ports and flows through the preform through its designed-in channels to facilitate wetting out. VARTM can produce parts with fiber volumes as high as 70% [Fabrication 2007].

Joint work between Krauss Maffei, Dieffenbacher, and Fraunhofer ICT [Graf et al. 2010] presents advancements in high pressure RTM, covering preforming, dosing systems, tool design, press, and machining. Figure 4.4 depicts RTM variants.

At the opposite extreme, Compression RTM combines high compressive pressure on the resin injected on the preform surface, thereby greatly reducing the time needed for impregnation. Compression RTM can produce composites with high fiber volume content while achieving high surface quality [Henning 2010]. Another variant of RTM is the separation of the process into an impregnation step and a curing step. The impregnation step can involve impregnation, preheating the preform, and precuring the resin. The cure step is conducted in the mold as a stand-alone process, thereby providing a short cycle time [Henning 2010]. Yet another RTM variant is High Pressure Injection RTM that injects the resin under high pressure (60-100 bar) to reduce the impregnation time, reduce void content, and enable high volume production [Henning 2010].

Resin film infusion (RFI) uses a high-viscosity resin film with a dry preform, and under the applied heat, vacuum, and pressure the resin is drawn into the preform. Resin films can be interleaved within the preform, enabling the use of high-viscosity and toughened resins to distribute uniformly. In 2006, Gurit Ltd. (UK) demonstrated that Class A panels could be made from carbon fiber/epoxy using its RFI process SPRINT [Dawson 2006a].

Vacuum infusion processing (VIP) uses vacuum to infuse the resin through dry layers of reinforcement. VIP can be used with all commercial fibers, cores, and resins with viscosity in the range of 50-1000 centipoise (cps).

4.10 Out-of-Autoclave Processing of Structural Components

Vacuum-bag-only processing (i.e., at atmospheric pressure) is well established for secondary structures, such as flaps and fairings in the aerospace industry. However, for primary structural components, less than 1% void content and autoclave-quality mechanical properties are required [Gardiner 2011]. The worldwide, cross-industry

Various Liquid composite molding processes

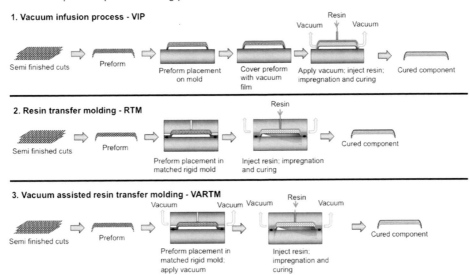

HP- RTM Process

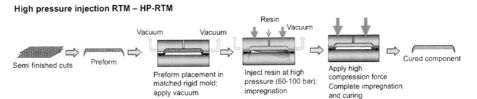

Advantages of High Pressure RTM:

- Rapid mold filling
- Improved impregnation quality
- Accelerated resin reactivity system can be applied – short cycle time
- Significant reduction of air entrapments and voids
- Excellent surface properties
- Low tolerance in thickness and 3D shape
- High process stability and repeatability
- Use of internal release agent – self cleaning system

Figure 4.4. Schematics of RTM variants [Graf et al. 2010].

(First published in the 2010 SPE ACCE, Troy, MI)

interest in out-of-autoclave (OOA) processing is driven by the need for faster cycle times as well as a desire to reduce energy consumption. Another advantage of OOA is reduced capital investment. These same drivers spurred the development and use of RTM and its variants and other liquid molding processes, including the latest compression molding of aerospace-grade thermoplastic composites [Gardiner 2011]. OOA prepreg systems are attractive for their potential to enable fast and affordable manufacturing. OOA prepregs have even resin distribution and can be cured at lower temperatures than autoclave cure [Gardiner 2011]. It has taken a number of years, but some OOA prepregs have reached physical property parity with autoclave-cure systems [Gardiner 2011]. However, many issues remain with OOA materials. A major issue is that, due to the time-dependent process of edge-breathing, cycle time might in fact be longer in order to achieve the required low void content for structural components [Gardiner 2011]. Surface quality, compatibility with adhesives, and automated layup also need to be addressed [Gardiner 2011].

4.11 Quickstep® Process

The Quickstep® process is an out-of-autoclave process that uses a heat transfer fluid to cure and join composites. Because the volumetric heat energy capacity of liquids is much greater than gases, the heat transfer to and from the laminate is much faster than in an autoclave. It is a balanced pressure and heated mold process that provides tight temperature control by circulating the fluid through the pressure chamber [Coenen et al. 2007]. A semi-rigid or free-floating rigid mold holds the composite laminate. Both laminate and mold are separated from the heat transfer fluid by a flexible membrane. The heat transfer fluid permits rapid heating to cure the laminate under low pressure and rapid cooling for demolding. In May 2011, Quickstep Holdings Ltd. (North Coogee, AU) announced its first commercial production-ready machine that provides more precise control over the ramp rates and a faster response to the exothermic reaction than its past machines, which have all been custom-built for research and development [Quickstep 2011].

4.12 Preforming Processes

Preforms are used in any liquid molding processes (RTM, RIM, wet compression molding) for mid- to high-volume closed molding production [Buckley 2008]. Net- or near net-shape preforms can be created in different ways, from spray forming to engineering fabrics formed with selective curing and/or stitching [Buckley 2008]. It is possible to incorporate cores, fasteners, and sensors into preforms.

The National Composites Center (NCC, Kettering, Ohio) has a computerized spray-up process called Large Scale Preformer (LSP) that robotically accomplishes the spray-up, compression, cure, cooling, and demold operations. The process uses powdered (dry) binders to avoid volatile emissions. The finished preform requires no trimming or special handling to avoid damage during molding operations [Dawson 2006b].

Materials Innovation Technologies, LLC (MIT-LLC) developed a three-dimensional engineered preform (3-DEP™) process. The 3-DEP™ process is an extension of pulp molding applied to reinforcing fibers as opposed to paper pulp. State-of-the art vacuum slurry technology is used to make pulp-molded preforms from essentially any type of chopped fiber — carbon, glass, natural fibers, or polymer fibers [Janney et al. 2009]. The process is capable of making very uniform parts with complex geometry. It also permits layering of different types of fibers through the thickness in desired geometric layups over the part. The preforms can then be molded into a composite using any liquid molding process such as vacuum infusion, RTM, liquid compression molding, etc. [Janney et al. 2009]. Initial work on their chopped carbon fiber thermoplastic composites is presented by Janney [2010].

4.13 Other Processes

A United States Air Force research program investigated a preform technology for oriented, discontinuous carbon fibers. The technology is called the Programmable Powdered Preform Process for Aerospace (P-4A) and was developed by Owens Corning [Reeve et al. 2000]. The technology consists of a computer-controlled chopper head mounted on a robotic arm to chop rovings and deposit fibers on a shaped screen to form a preform [Reeve et al. 2000]. The preform is then infused with resin to fabricate the composite part. Reeve et al. looked at the mechanical properties that could be attained with oriented, discontinuous carbon fiber/epoxy using P-4A compared to their baseline continuous fiber composite counterparts. They found that, compared to their continuous fiber counterparts, the P-4A composites retained more than 90% of the stiffness and 70–80% of the strength.

Takahashi et al. [2011] developed a low-pressure de-aeration process using vacuum molding to produce their CFCs from a knitted fabric prepreg. Dry layers, or air passages containing no resin, were created in the knitted fabric to allow air to escape. Using a vacuum pump during de-bulking of the stack and a tetra-functional powder epoxy integrated in the resin to prevent blockage of the air passages during manual layup were key to their success.

A surface finishing/compression molding process (SFC) developed by Valyi Institute for Plastic Forming (University of Massachusetts) combines resin extrusion, film finishing, and compression molding in one process [McCarthy et al. 2001]. The process consists of a finishing film placed over a mold cavity, resin extruded over the film from a "coat-hanger" die, mats of discontinuous fibers (glass and/or carbon) placed on top of the resin deposit, and closing the mold.

4.14 Tooling

Nickel Vapor Deposition (NVD) shell tooling provides conformal heating and cooling designs that provide uniform mold surface temperatures and shorter cycle times than conventional tools. The NVD process creates a solid, stress-free nickel shell over a CNC machined metal mandrel. Compared to conventionally used P20 tool steel, the NVD nickel shell provides similar hardness, a corrosion-resistant surface, and a three-fold increase in thermal conductivity [Sheppard 2007]. The technology can be applied to direct backfoam molds, spray molds, slush molds, RIM molds, or pressure molds.

Surface Generation Ltd. has developed a tooling technology to significantly reduce the machining of large billets of metal to create molds with complex shapes [Generation 2008]. Their technology is termed reconfigurable pin tooling, with two types: near-net-shape pin tooling (NPT) for low pressure and temperature molds, and subtractive pin tooling (SPT) for high pressure and temperature molds. In simple terms, multiple small blocks of metal are supported by pins. The positions of the pins are adjusted (in height) so the blocks of metal approximate the tool surface profile. The small blocks of metal are then CNC machined to the accurate surface profile. The resulting tool surface can be treated as required. The technology creates a tool that can be used for mid-volume production, with much less machining time and material waste.

Low Cost Tooling for Composites (LCTC), developed by Ciba Specialty Chemicals and Boeing, uses seamless epoxy patties to build layup tools. The epoxy compound is oven-cured and then CNC machined to produce lightweight, extremely accurate, dimensionally stable tools for fabricating prepreg prototypes and short-run parts in an autoclave [Industry 2000].

The fibretemp™ carbon fiber molds by fibretech composites (Germany) are a laminate structure in which the carbon fiber laminate provides both the loadbearing function and the resistance heating function. The laminate structure of the mold is similar to a heating panel, and the heating rate can be designed by structuring the layers [Funke 2009]. Fast heating rates, greater than 100°C/minute, are possible and temperatures of the mold can reach up to 150°C. The resistance heating requires little power, approximately 500 W/m².

Heated tools have been GKN's conventional approach to OOA curing because, whether electrically or fluid-heated, the autoclave and oven can be avoided [Gardiner 2011]. The higher cost of integrally heated tools is more than offset by the significant reduction in cycle time.

Graftech has studied the use of carbon foams (with densities of 0.16–0.32 g/cm³) as a core material for composite tooling [Kaschak et al. 2009]. Its working solutions include techniques for machining carbon foam, bonding foam blocks together to make larger tools, adhering uncured carbon fiber laminates to the foam surface, and drying and venting the material to remove unwanted moisture.

A study on the effect of various material and process parameters on the process-induced stress and deformation [Koteshwara and Raghavan 2001] considered, among other things, tool material and tool thermal mass by evaluating an aluminum high mass tool, an aluminum low mass tool, and an Invar tool. In an autoclave consolidation of carbon fiber/epoxy composite, the aluminum low mass tool resulted in maximum spring-back while an Invar tool resulted in the minimum spring-back. Such information is useful to the tool designer in terms of providing suitable allowances in the tool to achieve the desired final part dimensions and shape.

4.15 Closing Thoughts

As discussed in Chapter 3, both the aerospace and automotive industries are driving changes in carbon fiber composite technology to produce components that have lower material cost and targeted performance. Those developments will in turn lead to some developments in the manufacturing processes. For example, changes in processing temperatures can result in changes in tooling materials and heat sources, and changes in composite construction can lead to changes in material handling during component manufacture. Such future advancements driven by raw material and construction improvements will be additive to the advancements driven directly by the manufacturing process to improve areas such as part-to-part cycle time and energy efficiency.

References

Buckley, Daniel T. 2008. "High Volume Preforms for Structural Applications using Engineering Fabrics." SPE ACCE, Troy, MI (http://speautomotive.com/aca).

Buckley, Richard T., Jason T. Miwa, Rudolf H. Stanglmaier, and Donald W. Radford. 2006. "Light-Weight Composite Valve Development for High Performance Engines." SAE Paper No. 2006-01-3635. SAE International, Warrendale, PA. 2006.

Burkhart, Andrew and David Cramer. 2006. "Continuous-fibre reinforced thermoplastic tailored blanks." *JEC Magazine*, January-February 2006. no. 22.

Coenen, Victoria, Andrew Walker, Richard Day, Alan Nesbitt, Jin Zhang, and Bronwyn Fox, L. 2007. "The Effect of Cure Cycle Heating Rate on the Fibre/Matrix Interface." *Proceedings of the 28th Risø International Symposium on Materials Science: Interface Design of Polymer Matrix Composites – Mechanics, Chemistry, Modelling and Manufacturing.* Riso National Laboratory, Roskilde, Denmark. 2007.

Dawson, Donna. 2006a. "Carbon car hood: Class A and cost-effective." *High Performance Composites.* November 1, 2006. (http://www.compositesworld.com/articles/carbon-car-hood-class-a-and-cost-effective Accessed August 1, 2011).

———. 2006b. "Rapid Fiber Preforming On A Large Scale." *Composites Technology.* April 2006.

Dodworth, Antony. 2009. "Bentley Motors Develops Unique Directional Carbon Fibre Preforming Process for Chassis Rails." SPE ACCE, Troy, MI (http://speautomotive. com/aca).

du Toit, Werner, Willem Louw, Barry Crawford, and Pieter du Toit. 2010. "New Molding Process Offers Unique Levels of Design Complexity, Mechanical Strength, Cost Reduction for Long-Fiber Thermoplastic Composites." SPE ACCE, Troy, MI (http:// speautomotive.com/aca).

"Fabrication methods." 2007. *Composites World*. November 1, 2007. (http://www.compositesworld.com/articles/fabrication-methods Accessed August 1, 2011).

Feraboli, Paolo, Michael J. Graves, and Patrick B. Stickler. 2007. "Characterization of High Performance Short Carbon Fiber/Epoxy Systems: Effect of Fibre Length." SPE ACCE, Troy, MI (http://speautomotive.com/aca).

Funke, Herbert. 2009. "Electrically-heated moulds of CRP composite materials for Automotive Applications." SPE ACCE, Troy, MI (http://speautomotive.com/aca).

Gardiner, Ginger. 2010. "Aerospace-grade compression molding." *High Performance Composites*. June 30, 2010. (http://www.compositesworld.com/articles/aerospace-grade-compression-molding Accessed August 12, 2011).

———. 2011. "Performance requirements: CAI vsOHC." *High Performance Composites*. January 2011. (http://www.compositesworld.com/articles/performance-requirements-cai-vs-ohc Accessed August 12, 2011).

Generation, Surface. 2008. "'Goodbye Machining' Reconfigurable Pin Tooling." SPE ACCE, Troy, MI (http://speautomotive.com/aca).

Graf, Matthias, Erich Fries, Josef Renki, Frank Henning, Raman Chaudhari, and Bernd Thoma. 2010. "High Pressure Resin Transfer Molding - Process Advancements." SPE ACCE, Troy, MI (http://speautomotive.com/aca).

Grant, Carroll. 2005. "Automated Tape Layer Processing for Composite Components." SPE ACCE, Troy, MI (http://speautomotive.com/aca).

Henning, Frank. 2010. "Technology Development for Automotive Composite Part Production - New Materials & Processes." SPE ACCE, Troy, MI (http://speautomotive. com/aca).

"Industry Segment Profile - Composites." 2000. EPRI Center for Materials Production. no. 000000000001000135.

Janney, Mark A. 2010. "Initial Development of Thermoplastic Composites using Chopped Carbon Fiber." *Proc. Alabama Composites Conference*, August 25-26, 2010. Birmingham, AL.

Janney, Mark, Ervin Geiger, Junior, Tod Gunder, Neal Baitcher, and Roger Johnson. 2009. "Chopped Carbon Fiber Airplane Propeller Spinner." 41st International SAMPE Technical Conference - Witchita, KA. Oct 19-22, 2009.

Kaschak, David M., Richard Shao, Gary D. Shives, and Andrew J. Francis. 2009. "Machining, Bonding, Sealing, and Venting of Carbon Foam for Production Tooling." 41st International SAMPE Technical Conference - Witchita, KA. Oct 19-22, 2009.

Koteshwara, Madhava P. and J. Raghavan. 2001. "Parametric Study of Process Induced Deformation in Composite Laminates." SAMPE 2001 - Long Beach, CA. May 6 - 10, 2001.

McCarthy, Stephen, Qing Guan, Shawn McCarthy, Malar Rohith Shetty, Thomas Ellison, and Arthur Delusky. 2001. "Performance of Long Glass Fiber Reinforced Thermoplastic Automotive Part by Surface Finishing/Compression Molding Process." SAE Paper No. 2001-01-0442. SAE International, Warrendale, PA. 2001.

Norris, Robert E., Junior, Christopher J. Janke, Cliff Eberle, and Junior George E. Wrenn. 2000. "Electron Beam Curing of Composites Overview." SAE Paper No. 2000-01-1525. SAE International, Warrendale, PA. 2000.

Ouellette, Joseph. 2010. "Heatpipe / Thermosyphon Augmented Mandrels to Improve Cure Quality and to Reduce Cure Time in the Thermoset Pipe and Tube Filament Winding Process." SPE ACCE, Troy, MI (http://speautomotive.com/aca).

Potyra, Tobias, Dennis Schmidt, Frank Henning, and Peter Elsner. 2009. "Flexibility in the Direct Strand Moulding Compound (D-SMC) Process." SPE ACCE, Troy, MI (http://speautomotive.com/aca).

"Quickstep develops off-the-shelf production system." 2011. Composites World. June 13, 2011 (http://www.compositesworld.com/products/quickstep-develops-off-the-shelf-production-system Accessed June 27, 2011).

Reeve, Scott, Roger Rondeau, Gary Bond, and Fred Tervet. 2000. "Mechanical Property Translation in Oriented, Discontinuous Carbon Fiber Composites." SAMPE 2000 - Long Beach, CA. May 21 - 25, 2000.

Sheppard, Rob. 2007. "Nickel Vapor Deposition (NVD) Shell Tooling & Product Applications." SPE ACCE, Troy, MI (http://speautomotive.com/aca).

Smith, John R. and Uday K. Vaidya. 2010. "Progressive Forming of Thermoplastic Laminates." SPE ACCE, Troy, MI (http://speautomotive.com/aca).

Takahashi, Naoyuki, Yuji Kageyama, and Nobuya Kawamura. 2011. "Research of Multi-Axial Carbon Fiber Prepreg Material for Vehicle Body." SAE Paper No. 2011-01-0216. SAE International, Warrendale, PA. 2011.

"Teijin Receives Frost & Sullivan Award for Developing Lightweight CFRP for Automotives." 2011. *Omnexus by SpecialChem.* July 29, 2011. (http://www.omnexus.com/resources/print.aspx?id=28359 Accessed July 29, 2011).

Chapter Five

Machining and Joining

One must be entirely sensitive to the structure of the material that one is handling. One must yield to it in tiny details of execution, perhaps the handling of the surface or grain, and one must master it as a whole.
—Barbara Hepworth

5.1 Machining

CFC components are generally fabricated to near-net-shape, but often still require machining to trim the part to final dimensions or to drill holes for assembly. The inhomogeneous, anisotropic, and abrasive nature of CFCs creates challenges in retaining integrity of the material, with regard to delamination, pitting, heat damage, and burning. Tool wear exacerbates material damage that can be inflicted by machining, and considerable effort continues to be directed toward minimizing the rate of tool wear to increase production rates while maintaining acceptable cut quality. The automotive industry can benefit greatly from methods developed for the aircraft industry to meet its strict specifications on the quality of machined CFCs.

5.1.1 Tool Drilling

Most engineering structures contain holes and other openings for various purposes, and automotive structures are no different. Engineers know that any hole causes a stress concentration that needs to be accounted for in design. Different hole shapes create different magnitudes of stress concentration. The edge quality and integrity of the material surrounding the hole also affect the in-situ stress concentration, as well as crack initiation and fatigue life. Additionally, tolerances on the hole geometry are important for fastener installation and part assembly. These issues are common for all materials, but are exemplified with CFCs.

Lessons learned and developments made using CFCs in other industries such as aerospace and marine are quite applicable to the automotive industry. Regardless of the industry, joint quality depends on the performance and positional capabilities of the material removal and component fastening systems. Illustrating the technical complexity and challenges of achieving high-quality composite joints is an aircraft industry project called Automation for Drilling, Fastening, Assembly, Systems Integration, and Tooling (ADFAST) that was undertaken in the European Union with 10 partners beginning in 2001 and lasted more than three years. The goal of this project was to develop the technology to make automation of high-quality joining affordable. The target project deliverables included innovative orbital drilling equipment that would replace many existing operations, a new fastening system with improved control, novel low-cost reconfigurable tooling, and advanced machine control using high-precision metrology systems [Automation 2004]. Midway through the project, results on the orbital drilling were presented [Johansson, Ossbahr, and Harris 2002]. They compared conventional (dagger) drilling and orbital drilling with various drilling parameters, and evaluated the resulting effects on hole quality and mechanical properties of AS4/8552, a unidirectional carbon/epoxy pre-pregged composite. They found that specimens with orbitally drilled plain holes had higher fatigue life than those with conventionally drilled holes. The opposite was true for countersunk holes but may have been caused by the higher surface roughness in the countersunk region due to the high feed rate in the orbital drilling. The intricacies of quality hole making were not yet solved. At the conclusion of the project, SAAB Aerostructures, one of the project partners, reported that a demonstrator (prototype) of a recon-figurable, automatic-based jig was developed that they were looking to improve and optimize [SAAB 2004].

The orbital drilling process was developed by Novator (Sweden) [Destefani 2011]. Orbital drilling simultaneously machines material both axially and radially by rotating the cutting tool about its own axis and about a principal axis offset from the tool axis. A hole is produced by moving the tool in both directions. A single tool can produce holes of different diameters by adjusting the offset of the principal axis. When drilling holes, this simultaneous axial-radial movement reduces the axial force, thereby reducing the risk for delamination in composite laminates. Cutting forces in orbital drilling are as low as one-tenth of the forces generated in conventional drilling using a spiral cutter [Kihlman 2005]. The SAAB Ericcson Space Board application of orbital drilling proved that it could drill a 0.8-mm-thick composite laminate without delamination, which was an extreme, challenging case [Kihlman 2005]. Another advantage over conventional drilling includes more efficient heat and chip extraction because the tool diameter is less than the hole diameter, and the cutting edge is in intermittent contact with the hole edge. Additionally, the cutting tools do not drift when entering curved, inclined, or irregular surfaces. Because the orbital motion is strictly mechanical, the drilling can be done much faster and with higher accuracy. Orbital drilling produces burr-free holes in metals and delamination-free holes in composites, and permits dry chip removal because there is less heat buildup [Waurzyniak 2002]. An additional feature of orbital drilling is that it can machine features, such as a cavity, by moving the tool radially [Destefani 2011]. Novator's CNC Orbital Drilling unit is shown in Fig. 5.1.

Figure 5.1. Orbital drilling unit by Novator
[Destefani 2011; Larsson 2009].

In an interesting twist, Kihlman [2005] looked to the highly automated spot welding operations in the automotive industry for insight into increasing the automation in aerospace joining. Industrial robots used for spot welding in the automotive industry have good repetitive accuracy, but their absolute accuracy is 10-15 times less. In general, for car assembly a spot weld gun is positioned within ±1.2 mm, whereas a drilling machine in aircraft assembly requires positioning within ±0.2 mm or better [Kihlman 2005]. Typical automation machines in aerospace for high accuracy must have high dynamic stiffness, and tend to be large, expensive, and dedicated to a narrow scope [Kihlman 2005]. The object of Kihlman's work, in part, was to determine if and how industrial robots, such as those used in automotive, can be utilized to enable less expensive aircraft assembly automation. It was found that off-the-shelf industrial robots could not cope with forces other than gravitational loads, and thus low forces in the robotic process are critical to achieve high accuracy. When cutting a stack of materials, the air gaps between the layers of materials need to be removed to prevent burrs from entering between the layers. Although the forces in orbital drilling are 8–10 times less than those in conventional drilling, the pressure needed to remove the air gaps between stacked materials means the forces on the robot will not simply be gravitational, and deflections will be generated in the (industrial) robot. Kihlman recommends using a force cone concept or a metrology system to detect the robot deviation and compensate accordingly. Today, industrial robots are calibrated using metrology systems, but the calibration is essentially static [Kihlman 2005]. Situations such as drilling are dynamic, not static, and furthermore involve contact scenarios such as engaging a surface prior to drilling. Metrology-integrated robot control was introduced as a solution by Kihlman, and with the equipment used in his research a feedback loop of 3–5 seconds enabled the industrial robots to reach and maintain high accuracy.

In the aerospace industry, there is an increasing use of the stack materials — either various layers of composites, composite/metal layers, or foam or other core material wrapped in composites. Stack thicknesses vary from less than a quarter of an inch to several inches. Each Boeing 787 Dreamliner contains approximately 35 tons of CFCs, and a large percentage is in a stacked form with aluminum or titanium alloys [Destefani 2011]. The automotive industry may never see stacks several inches thick, but it will see composite/metal layers at joint interfaces. Drilling through different material types or constructions can cause fraying, delamination, or changes in hole size. Examples of potential quality issues in drilling composite/metal stacks are shown by Destefani [2011].

To drill layers of different types of materials requires the right technology. Even layers of different types of composite materials, or different constructions of CFCs, require the right drilling technology. The optimal process is usually a compromise between reducing the cutting force, to minimize composite delamination, and extending the tool life [Destefani 2011].

The construction of the CFC (unidirectionals at different orientations, weaves, braids, etc.) or stack, the thickness of the material or stack, which layer is penetrated first and which layer is on the backside, and whether or not there is a backing material all affect the drilling parameters and tools. For thick CFCs, in which a lot of heat is built up during drilling, a cutting tool with narrower flutes, wider gullets, and tighter spirals allows faster penetration. (Coolant is not generally used when drilling CFCs because CFCs can absorb the coolant and swell.) If aluminum is on the backside of a CFC, a high-shear drill with a sharp point angle minimizes burr height of the exit hole. However, if CFC is on the backside, a different feed rate and a drill/reamer combination tool with multifaceted drill points works well [Destefani 2011].

The cutting tool and speed preferred for a CNC drilling operation are often different from that which would be preferred for manual drilling. It is recommended that the end user work closely with the cutting and tool supplier to select the correct drill for their application.

5.1.2 Tool Wear Compensation

Because of the abrasiveness, inhomogeneity, and anisotropy of CFCs, tool wear is an issue for all mechanical drilling methods. Some of the effect of tool wear is a reduction in hole diameter as the number of holes drilled increases. Figure 5.2, an example from Larsson, Eriksson, and Rydberg [2009], illustrates this for orbital drilling a 9/16" (14.3 mm) hole in CFC over a run of 300 holes.

Larsson, Eriksson, and Rydberg undertook a study to develop a reliable and predictable compensation algorithm that could be applied to the orbital drilling process to maintain hole dimension and roundness throughout a series of holes. The algorithm they developed factored in cutting tool dimensions, the current wear of the tool (in terms of current compensation needed to produce the correct hole diameter), and the material being drilled. Each material being drilled has to have its own wear factors and

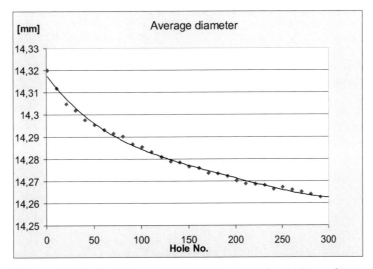

Figure 5.2. Typical change in hole diameter for drilling of CFC
[Larsson, Eriksson, and Rydberg 2009].

compensation curves determined. Once those are established and stored in a database, the orbital drilling compensation algorithm can be implemented. Results presented [Larsson, Eriksson, and Rydberg 2009] show that orbital drilling with tool wear compensation can produce a large number of holes with tight tolerances and significantly increase the life of the tool.

5.1.3 Tool Cutting/Trimming

Although CFC components are generally manufactured to near-net shape, additional machining such as trimming of the edges to attain final dimensions and edge profile is often conducted. Because CFCs are abrasive, inhomogeneous, and anisotropic, the machining usually removes material through fracture and fragmentation as opposed to continuous chips generated with metal [Snider 2010; Sheikh-Ahmad and Sridhar 2002]. Mechanical cutting of CFCs can also cause internal voids due to fiber pullout, and also creates a heat-affected zone that can melt the resin and induce a recast state with microcracks [Snider 2010]. As with drilling, a major concern with edge trimming of CFCs is tool wear, both in terms of tool replacement and the quality of machining. Inadequate tool sharpness can lead to delamination, pitting, and burning [Sheikh-Ahmad and Sridhar 2002]. Maintaining cut quality at desired production rates requires cutting tool materials such as polycrystalline diamond (PCD) that provide good surface quality and long tool life.

PCD tools are very expensive, but since the development of chemical vapor deposition (CVD), commercial diamond-coated tungsten carbide tools provide an alternative. The CVD diamond-coated tools combine the extreme hardness of the coating with the toughness of the tungsten carbide substrate, along with the design flexibility of insert tooling [Sheikh-Ahmad and Sridhar 2002]. The process methods involved in CVD diamond film growth on tungsten carbide tool surfaces and the unique properties of diamond thin films is nicely presented by Engdahl [2006]. In the study by Sheikh-Ahmad and Sridhar [2002], it was concluded that diamond-coated tools are best suited for finish cutting of CFCs, and a better surface roughness is generated at low feed speed.

5.1.4 Abrasive Waterjet Cutting

There are several alternatives to traditional (mechanical) machining of materials in general, including laser, electrical discharge machining (EDM), ion beam/electron beam cutting, and microwave cutting. When it comes to machining CFCs specifically, these methods have some disadvantages. Lasers can easily introduce a heat-affected zone because carbon fiber requires high heat to be cut. EDM is a challenge for materials with poor conductivity such as glass fiber and ceramic, which may exist in a carbon fiber composite layup with multiple materials. Ion beam/electron beam and microwave are generally limited to thin sections, and cannot cut contoured surfaces with tight tolerances [Lui et al. 2010]. Nanomaterials, which are stronger and tougher than carbon fibers and whose properties can change from conductive to nonconductive and/or from nonreflective to reflective with sharp gradients, are emerging as additions to more conventional CFCs, and will certainly require nonmechanical machining.

Abrasive waterjet (AWJ) cutting can have some advantages over conventional cutting operations such as hand routers, NC mills, and waterjets. Some of these advantages are AWJ components do not wear as fast as traditional cutting bits, the abrasive media used is cheaper than traditional bits, and AWJ trimming uses less tooling and fixturing than traditional trimming [Krajca and Ramulu 2001]. AWJ cutting speed is also typically twice as fast as mechanical cutting [Snider 2010]. Because there is no heat-affected zone, the integrity of the material's structure and chemistry is retained [Lui et al. 2010].

In 2001, folks at Boeing and the University of Washington reported on a study that examined the use of abrasive waterjet piercing of CFC and the resulting effects on hole quality of carbon/epoxy composites [Krajca and Ramulu 2001]. They found that increases in standoff distance (from nozzle to composite) resulted in hole taper ratio increases, delamination area and jet entrance delamination decreases, and initial damage zone increases in width and depth. The waterjet pressure also significantly affected hole quality; increases in pressure resulted in taper ratio increases, delamination area and jet entrance and exit delamination increases, and an increase in percentage of plies delaminated.

For machining hygroscopic materials, abrasive cryogenic jets (ACJ) using liquefied nitrogen (LN_2) can be used. It was discovered that ACJ could mitigate piercing damage in a variety of delicate materials [Lui et al. 2010]. Most of the LN_2 evaporates before entering into a blind hole, and thus pressure buildup inside the blind hole is reduced, preserving the material integrity. However, ACJ is not a viable tool for industry because it is bulky, potentially hazardous to the operators, and not economical. Lui et al. [2010] therefore developed a novel flash AWJ, or FAWJ, that superheats the high-pressure water such that the superheated water changes to steam upon exiting the orifice and mixing tube. They demonstrated the superiority of this FAWJ over AWJ in preserving the structural integrity of machined composites when piercing delicate materials.

The use of AWJ with robots as waterjet tool manipulators was discussed by Snider [2010]. Five-axis gantry robots are designed with a high degree of structural stiffness to achieve the desired cutting accuracy. The units are usually installed in a controlled environment and on a machine tool foundation with helical rack and precision gear boxes, along with closed loop control systems, to provide smooth motion and high accuracy. The linear axes are compensated using laser interferometers, and linear encoders are used to improve accuracy on large-scale systems. AWJ with five-axis robots can achieve accuracies of ± 0.002 in (0.05 mm). Six-axis robots are mass-produced and highly repeatable, although not accurate until they are compensated, usually using a teach pendant [Snider 2010]. Rigid six-axis robots are seeing more use now that their accuracies have reached ± 0.010 in (0.25 mm) and better [Snider 2010].

5.1.5 Laser Cutting

Laser cutting has been looked at for cutting CFCs. The cut quality obtained by common CO_2 lasers can vary greatly and depends on the particular construction of the CFC. CFCs containing large amounts of epoxy resin (i.e., lower fiber volume) cut more slowly and have more dark discoloration and residue than CFCs with higher fiber volumes. In an application by Synrad [Synrad 2010], a carbon fiber acrylic composite 0.1 in (2.5 mm) thick was laser cut using a 200-W laser with a 2.5-in focal length lens that produces a 0.004-in (0.1-mm) focused spot size. A 50-psi air assist was used to shield the lens and reduce flame-up of the CFC. A cut rate of 80 in/min (2 m/min) was achieved and the cut edges had only slight discoloration. Although avoiding flame-up of CFCs while achieving a good quality cut and fast rate is a challenge, other manufacturers of laser cutting machines identify the capability to laser cut CFCs [Laser Cutting Systems; Laser Cutting of Other Materials].

5.2 Joining

The challenges of joining CFCs continue to get lots of attention from material suppliers, composite manufacturers, and end users. As with machining of CFCs, the joining of CFCs in the automotive industry can adopt technologies and lessons learned in aerospace and marine. One of the most critical aspects of using CFCs in a vehicle is how they will be attached to other vehicle components. When selecting an attachment or joining method, the performance of the joint as well as the cost and time to produce are key. Because holes and mechanical fasteners create discontinuities in the CFCs — namely breaching the reinforcement — other methods are often used where appropriate. These include adhesive bonding and co-curing, due to the fact that, to date, most structural CFCs are based on thermoset resins. However, for critical joints in aerospace, the joining process is usually augmented with mechanical fasteners.

5.2.1 Bonding

5.2.1.1 Adhesive Bonding

Adhesive bonding has been used for decades in aerospace, and has a decent history in automotive too, particularly in low-volume niche vehicles. The extruded aluminum space-frame, carbon fiber composite transmission tunnel, and energy-absorbing crash structures of the 2002 Aston Martin Vanquish were all adhesively bonded together [Hill 2003]. Of course, different material combinations required different types of adhesives. The front-end crash assembly of the Vanquish was made of carbon fiber/glass fiber-epoxy using RTM, the A-pillars were made from braided carbon fiber over polyurethane core, and the transmission tunnel was made from RTM CFC. The CFC components were bonded using a medium-level, two-component polyurethane adhesive. This choice was made after considerable laboratory testing, and accounting for the ambient conditions of the assembly plant, the time needed to manually dispense the adhesive and clean up the joint, and the amount of cure needed before the framing jig was removed for the next part.

It was reported in 2006 that the limited-edition BMW M3 CSL sedans also used a two-part polyurethane adhesive to bond its carbon fiber/epoxy roofs to the vehicle [Technology 2006]. For installation, the vehicle was taken off the production line for workers to manually clean the roof and roughen the mating surfaces. A continuous bead of adhesive was manually placed into the steel roof opening, and the roof was put in place, pressure applied, and cured at room temperature for 100 minutes.

There were only 1500 limited-edition M3s. For larger volume production, robotic application of the adhesive bead is typical. Parts are robotically placed in the bonding cell, aligned, and secured to achieve gauge dimensions and tolerances (GD&T) of the finished assembly, and the adhesive bead is robotically placed, with or without hot air impingement at the adhesive region to accelerate the cure. A schematic of such a system is shown in Fig. 5.3.

The most common adhesives for automotive CFCs are polyurethanes and epoxies, but acrylates or phenolics can also be used. Epoxies typically have higher shear modulus than polyurethanes, but the higher elasticity of the polyurethanes often dominates performance. Adhesives go through rigorous testing to qualify to OEM material and engineering specifications. Depending on the component, validation and/or qualification will include static and fatigue lap shear, static and dynamic torsion, peel tests, paint evaluations, and crash tests. The adhesive must have sufficient elastic elongation to handle road loads, and often must accommodate dissimilar materials that have quite different coefficients of thermal expansion. Qualification tests also typically include environmental exposure and aging; bonded specimens are exposed to a variety of cyclic temperature and environmental conditions, which can consist of hot, cold, water soak, humidity, and salt spray. When tested to failure, adhesive bonds are to fail cohesively and demonstrate tear in the substrate.

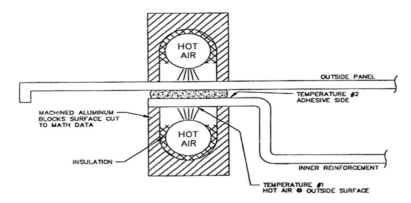

Figure 5.3. Schematic of hot air adhesive bonding.
(Image courtesy of D. Schilkey, Alliance Engineering Inc., Rochester Hills, MI)

5.2.1.2 Microwave and Electron Beam Adhesive Bonding

Adhesives that cure under radiation include electron beam (EB) curable and ultraviolet (UV) curable. Their advantages include fast cure, reduced energy consumption, and no residual stresses. UV-curable adhesives have been successfully implemented because of their rapid cure rate and their compatibility with automatic dispensing equipment. For some systems, microwave adhesive bonding can produce stronger bonds.

Compared to the adhesive, the composite parts being bonded have low loss factors. Therefore, even though the substrate parts can be thick, the microwave energy is directly transferred into the interface and cures the adhesive rapidly [Zhou et al. 2001]. Additionally, bonds produced by microwaves have mechanical and physical properties equal to or better than those of thermally produced bonds [Zhou et al. 2001].

In contrast to UV curable adhesives, EB curable adhesives have been slow to develop due to a lack of commercial applications [Byrne et al. 2000]. EB curable adhesives have potential to replace thermally cured adhesives for bonding large structures in fewer steps and reduced cost compared to autoclave-cured adhesive bonds [Byrne et al. 2000].

5.2.1.3 Welding/Weld Bonding

Laser welding, hybrid laser welding, and weld bonding have been explored in more recent years as a means to achieve very rapid, high quality joining of CFCs and CFC to metals.

Laser welding has been demonstrated and successfully applied to polymeric materials for a number of years, with laser manufacturers touting such capabilities, e.g., SPI [2011] and Olowinsky [2011]. However, CFCs, with their carbon fibers, bring challenges in managing the heat intensity. Applying laser welding to CFCs is currently still in the research and development phase, with the author being one who is working in this area.

Hybrid laser welding involves a supplemental radiation to the laser welding process. In work by Geiger et al. [2008], polychromatic radiation emitted by halogen lamps provided improvements above conventional laser welding of black and natural polycarbonate (PC), and colored polymethylmethacrylate (PMMA) to clear or colored PC. The secondary radiation from the halogen lamps is pooled with the primary radiation from a diode laser to one focal point using optical fibers and lenses. The hybrid process allows the energy being absorbed by both parts, not just the infrared absorbing part, to be controlled. This creates a larger weld processing window that provides a higher gap-binding capability and lower residual stresses at the weld seam. Additionally, the feed rate can be increased, 2–5 times in their application, while attaining equal or better tensile strength. This hybrid approach should also aid in welding large three-dimensional parts.

Weld bonding is another hybrid type of joining, in which a bonding material is used at the joint interface to provide compatibility between dissimilar materials, and then the system is welded together.

5.2.2 Mechanical Fasteners

Although advances in other fastening methods will result in their increasing implementation, there will always be a need for joining methods in areas where a need for repair or replacement is foreseen (e.g., liftgates), and these are best served by mechanical fasteners. As a result, efforts will continue to improve mechanical fasteners too, whether for greater automation, easier assembly, weight savings, cost savings, or improved performance.

As structural CFCs gain greater implementation in the automotive industry, there will be cases where the composite component is joined to another composite component and cases where the composite component is joined to a metallic one. This scenario also exists in aerospace. If the metallic component is on the backside or blind side of the composite component, one can use standard blind, metallic fasteners that would be used for metallic assemblies [Keener and Luhm 2004]. However, for joints of all composite components or for joints with the composite component on the blind side of a metallic component, the fasteners must be properly chosen and designed for the joint components' material characteristics.

The mechanical fastener design must accomplish several tasks: it must distribute surface loads uniformly to avoid stress concentrations; it must create high clamp loads with minimal pressure on the composite surfaces; it must prevent movement of the fastener elements when in use in the vehicle; and it must avoid galvanic interaction with the CFC. Table 5.1, developed by Keener and Luhm [2004], lists different mechanical fastener types that are most commonly chosen to join certain composite materials. Some of the recommendations for performance are also given.

Solid rivets, threaded pins, two-piece bolts, and blind fasteners made of titanium, stainless steel, and aluminum alloy materials are all used in the aerospace industry to join composites, but titanium and Monel are used the most to resist galvanic corrosion with CFCs. Nickel-base stainless steel is also compatible with CFCs, but the highest compatibility is found with fasteners made from composite materials. Conventional fasteners can be pre-coated to guard against corrosion, or standard aluminum or steel fasteners can be insulated with glass fiber or adhesive scrim [Keener and Luhm 2004]. Much effort has been put into developing composite fasteners in recent years to provide complete material compatibility.

Another consideration favoring composite fasteners is thermal expansion. Aluminum and stainless steel have thermal expansion coefficients far greater than that of CFCs. In service, large differences in thermal expansion between the fastener and the composite

Table 5.1. Commonly Used Mechanical Fasteners

Fastener			Recommended Applications			
Type	Mat'l	Finish	Epoxy/ Graphite	Kevlar	Fiber glass	Honey comb
Blind Rivet [1]	5056 Al	None	NR	E [8]	E [8]	
	Monel	None	G [8]	E [8]	E [8]	
	A-286	Passivated	G [8]	E [8]	E [8]	(5)
Blind Bolt [2]	A-286	Passivated	E [8]	E [8]	E	
	Alloy Steel	Cadmium	NR	E [8]	E	
Pull-type Lockbolt	Titanium	None	E [4]	E [3]	E [3]	G or NR [6]
Pull-type Lockbolt	7075 Al	Anodized	NR	E	E	NR
Stump-type Lockbolt	Titanium	None	E [4]	E [3]	E [3]	G or NR [6]
Asp® Fastener	Alloy Steel	Cad/Nickel	G [7]	E	E	E

Notation: **NR** = Not Recommended, **E** = Excellent, **G** = Good

Notes:
1) Blind rivets with controlled shank expansion.
2) Blind bolts are not shank expanding.
3) Fasteners can be used with flanged titanium collars or standard aluminum collars.
4) Use flanged titanium collars.
5) Performance in honeycomb structure should be substantiated by installation testing.
6) Depending upon fastener design.
7) Nickel plated Asp fasteners only.
8) Metallic structure required on backside of stack-up.

component can add significant stresses to the joint and adversely affect the preload or clamping load, which would directly impact the performance of the joint in fatigue [Keener and Luhm 2004].

When selecting or designing a mechanical fastener, consideration must also be given to creep of the fastener hole and long-term compression.

Critical parameters in composite joints are tensile pull-through, static lap shear, and lap-shear fatigue [Keener and Luhm 2004]. Even though fasteners composed of composite materials will have relatively low shear strength compared to their metallic counterparts, the levels are still acceptable for composite structural applications. This is because the allowable bearing stresses in the parent composite material are also low, meaning that the full shear capability of standard metallic fasteners is seldom developed when joining composites [Keener and Luhm 2004].

Fastener type and head configuration affect the overall joint integrity, including failure modes and failure loads. Because composites have lower compression strength than metallic materials, large head configurations are needed to distribute the load over a large area. Head styles are typically 100° and 130° flush or protruding. Interference-fit installations can create delamination if the interference is too high, and clearance-fit installations can create fretting if the fit is not close enough. There are special sleeved fasteners that provide net-fit installation while limiting damage in clearance-fit installations, and adhesives are sometimes used to reduce fretting.

There are other manufacturing and assembly issues with mechanical joining of composites. The matrix can be damaged from concentrated loads applied during drilling and installation. Delamination can also be created during drilling and installation, as well as in operation. Because mechanically fastened composite joints are designed to fail in bearing, delamination is a much more serious defect than fiber breakout [Keener and Luhm 2004]. Water can intrude between the fastener and the composite structure, or exposed carbon fibers can absorb water, which weakens the structure. Arcing between fasteners could occur if electrical continuity between the composite fibers and the fastener is not maintained. A number of very practical joint design recommendations are provided by Keener and Luhm [2004].

Rivets
The solid rivet is the simplest fastener type. The conventional solid metallic rivet such as titanium or Monel tends to radially expand in the hole during installation and can induce separation or peeling damage of composite laminates. Solid composite rivets can minimize or eliminate the issues of solid metallic rivets [Keener and Luhm 2004]. Threaded fasteners made from fiber-reinforced linear aromatic polyetheretherketone (PEEK™) are being used in aerospace applications with success [Keener and Luhm 2004].

A riveting system developed by MTorres and used to install rivets on the trailing edges of the HTP elevators in most Airbus aircraft is described by Marin et al. [2002]. The elevators are made of stacks of aluminum / CFC / aluminum sandwiches, or a combination of stainless steel / CFC / stainless steel and aluminum / CFC / aluminum sandwiches, with stack thicknesses ranging from 7.2–12.8 mm. The constructions and thicknesses are not likely to be reproduced in automotive applications, but the general operating principles of the riveting system could. The system contains a reconfigurable flexible tool that consists of seven retractable positioners and 20 numerically controlled (NC) actuators to accommodate parts of different sizes. The positioners are pneumatically actuated fixtures that hard index the part in the working envelope of the machine, and they are retracted during part measurement. Each of the 20 NC actuators can be driven in the Z direction to modify its height to accommodate the geometry of the part to be riveted. A vacuum cup on top of each actuator creates an air cushion that eases the handling and positioning of the part in the working envelope of the system, and it firmly holds the part in place during riveting. The riveting machine is fully NC driven, and its automated operations include measurement of part alignment, drilling, upper and lower countersinking, grip length measurement, rivet selection and feed, sealant application, riveting, and measure and inspection of part waviness [Marin et al. 2002].

Other operational features include drill bit breakage sensor, automatic countersinking height adjustment, drilling tool speed adjustment, riveting force limitation, nosepiece clamping force adjustment, and selection for automatic operation in one of three modes — drilling only, complete drill/rivet cycle, or riveting only.

5.3 Closing Thoughts

One of the most challenging aspects of implementing CFC components in vehicle design is attaching them to the rest of the vehicle. This usually requires machining and joining, which must be done in a manner that retains the mechanical properties of the CFC component as well as possible, provides a strong and durable joint, is cost-effective, and fits with the OEM assembly process and vehicle production rate.

References

"Automation for drilling, fastening, assembly, systems integration, and tooling." 2004. *CORDIS simple search* (http://cordis.europa.eu/search/index.cfm?fuseaction=proj. document&PJ_RCN=4965803 Accessed August 20, 2011).

Byrne, Catherine A., Daniel L. Goodman, Giuseppe R. Palmese, James M. Sands, and Steven H. McKnight. 2000. "Electron Beam Curable Adhesives for Out-of-Autoclave Bonding of Large Composite Structures." SAMPE 2000 - Long Beach, CA May 21 - 25, 2000.

Destefani, Jim. 2011. "Hole in Four ...or More." *Cutting Tool Engineering, Volume 63, Issue 1* (http://www.ctemag.com/aa_pages/2011/110102-Drilling.html Accessed August 20, 2011).

Engdahl, Niels Chris. 2006. "CVD Diamond Coated Rotating Tools for Composite Machining." SAE Paper No. 2006-01-3153. SAE International, Warrendale, PA. 2006.

Garrick, Richard. 2007. "Drilling Advanced Aircraft Structures with PCD (Poly-Crystalline Diamond) Drills." SAE Paper No. 2007-01-3893. SAE International, Warrendale, PA. 2007.

Geiger, Rene, Oliver Brandymayer, Frank Brunnecker, and Chris Korson. 2008. "Hybrid Laser Welding of Polymers." SPE ACCE, Troy, MI (http://speautomotive.com/aca).

Hill, John. 2003. "Adhesively Bonded Structural Composites for Aston Martin Vehicles." SPE ACCE, Troy, MI (http://speautomotive.com/aca).

Johansson, Sten Å. H., Gilbert C. R. Ossbahr, and Tom Harris. 2002. "A Study of the Influence of Drilling Method and Hole Quality on Static Strength and Fatigue Life of Carbon Fiber Reinforced Plastic Aircraft Material." SAE Paper No. 2002-01-2650. SAE International, Warrendale, PA. 2002.

Keener, Steven G. and Ralph R. Luhm. 2004. "Development of Non-Metallic Fastener Designs for Advanced Technology Structural Applications." SAE Paper No. 2004-01-2821. SAE International, Warrendale, PA. 2004.

Kihlman, Henrik. 2005. "Affordable Automation for Airframe Assembly - Development of Key Enabling Technologies." *Linkoping Studies in Science and Technology* no. Thesis No. 953 (http://130.236.35.10/ps/staff/henki/docs/Affordable_Automation.pdf Accessed August 20, 2011).

Krajca, Scott E. and M. Ramulu. 2001. "Abrasive Waterjet Piercing of Holes in Carbon Fiber Reinforced Plastic Laminate." 33rd International SAMPE Technical Conference - Seattle, WA - November 5 - 8, 2001.

Larsson, Eskil, David Eriksson, and Patrick Rydberg. 2009. "Tool Wear Compensation." SAE Paper No. 2009-01-3216. SAE International, Warrendale, PA. 2009.

"Laser Cutting of Other Materials." (http://www.rofin.com/index.php?id=493&L=1 Accessed October 8, 2011).

"Laser Cutting Systems." (http://www.elengroup.com/laser-cutting/ Accessed October 8, 2011).

Lui, H.-T., E. Schubert, D. McNeil, and K. Soo. 2010. "Applications of Abrasive-Fluidjets for Precision Machining of Composites." SAMPE 2010 - Seattle, WA May 17-20, 2010.

Marin, Diego Perez, Juan Astorga, Jose Morazo, Jose Ramallo, Rufino Jimenez, Juan Francisco Morales, and Francisco Giménez. 2002. "A Versatile Riveting System for Metal/CFC Structures." SAE Paper No. 2002-01-2644. SAE International, Warrendale, PA. 2002.

Olowinsky, Alexander. "Lasers In Plastics Technology." Fraunhofer Institute for Laser Technology ILT (http://www.ilt.fraunhofer.de/eng/ilt/pdf/eng/products/Lasers_in_Plastics_Technology.pdf Accessed August 22, 2011).

———. "Micro Joining with Laser Radiation." Fraunhofer Institute for Laser Technology ILT (http://www.ilt.fraunhofer.de/eng/ilt/pdf/eng/products/Micro_Joining_with_Lasers.pdf Accessed June 13, 2011).

SAAB. 2004. "Automated assembly in the future." *SAAB Company News Brief - 3/2004* (http://www2.saabgroup.com/SaabGroup.Web.WebSite/Templates/AdvancedContentPage.aspx?NRMODE=Published&NRNODEGUID={53A0B611-1C11-4C3B-A628-B95749D8C93F}&NRORIGINALURL=%2Fen%2FAboutSaab%2FOrganis ation%2FSaabAerostructures%2Fnewsbrief%2F200403%2Fautomated_assembly. htm&NRCACHEHINT=Guest Accessed August 20, 2011).

Sheikh-Ahmad, J. and G. Sridhar. 2002. "Edge Trimming of CFRP Composites with Diamond Coated Tools: Edge Wear and Surface Characteristics." SAE Paper No. 2002-01-1526. SAE International, Warrendale, PA. 2002.

Snider, Duane. 2010. "Precision Waterjet Cutting in the Composites Industry Utilizing Robots for High Quality Accurate Machining." SPE ACCE, Troy, MI (http://speautomotive.com/aca).

SPI. "Fiber Laser Welding of Plastics." (http://www.spilasers.com/Applications/red-POWER/Fiber_Laser_Welding_of_Plastics_.aspx? Accessed June 13, 2011).

Synrad. 2010. "Laser Cutting Carbon Fiber Reinforced Plastic Composites." *Newsletter 245, September 2, 2010.* (http://www.synrad.com/e-newsletters/09_02_10.htm Accessed August 22, 2011).

Technology, Composites. "BMWs Carbon Fiber Roof Attached With Polyurethane Adhesive." *Composites World February 1, 2006* (http://www.compositesworld.com/articles/bmws-carbon-fiber-roof-attached-with-polyurethane-adhesive Accessed August 22, 2011).

Waurzyniak, Patrick. 2002. "Holemaking with Precision." *Manufacturing Engineering* no. November 2002 Vol. 129 No.5. (http://sme.org/cgi-bin/find-articles.pl?&02nom003&ME&20021105&&SME& Accessed August 20, 2011).

Zhou, Shuangjie., Hua Yang, Liming Zong, and Martin C. Hawley. 2001. "Status of Microwave Adhesive Bonding Research." SPE ACCE, Troy, MI (http://speautomotive.com/aca).

Chapter Six

Reclaiming/Recycling Carbon Fiber Composites

In the long term, economic sustainability depends on ecological sustainability.
—America's Living Oceans

6.1 Introduction

Each year about 10–15 million vehicles reach their end of life in the U.S. alone [Jody et al. 2009]. Currently, about 75% of the weight of the vehicle is metals content, which can be recycled. Concerted efforts to increase the CFC content in vehicles to achieve lightweighting goals are occurring simultaneously with the call for recycling even more vehicle content, and the recycling of CFCs is far less developed than that of metals.

The average useful life of a vehicle is about 13 years, and about 95% of retired vehicles are processed by dismantlers and shredders [Jody et al. 2009]. There are more than 15,000 dismantlers in the U.S. that recover the automotive fluids and remove parts for remanufacturing or resale, and then sell the rest of the vehicle remains to a shredder. Shredders sell the recovered metallic material while the non-metallic material becomes shredder residue [Jody et al. 2009]. Of the almost 5 million tons of shredder residue generated each year, about 30% is polymer based [Jody et al. 2009].

Shredder residue is a complex mixture of a small amount (about 10%) of metals and roughly equal amounts of organic and inorganic material. The residue and particularly the mixture of various polymers are difficult to separate. Additionally, any CFCs reaching the shredder would lose their continuous or long lengths of fibers upon which they attain their valuable mechanical properties. For CFCs, the value of the recovered material will be greater if the recovery process keeps that material segregated and more intact than the traditional vehicle shredding process allows.

On the surface, recovering and reusing or recycling materials from retired vehicles should reduce energy usage while providing a new material stream for automotive or other applications. However, for many polymeric materials, CFCs included, the economics are not yet established. The economics of recycling depends on the cost to recover the material from the waste stream and process (clean and purify) it for reuse, and the value of the recovered material [Jody et al. 2009].

The majority of CFCs and other composite materials are currently discarded in landfills [Connor 2008]. In some locales, CFCs are classified as hazardous waste and create significant burdens relating to their disposal. Due to legislative and environmental concerns, there is a global effort to begin reusing CFCs and other reinforced plastics, and to reclaim the costly reinforcing fibers themselves. This includes manufacturing waste in both pre- and post-cure states, as well as post-consumer waste. Today, the largest source of CFC waste is from manufacturing scrap, including expired prepreg, prepreg offal, part trimmings, and off-specification parts. In the aviation section, it is estimated that the amount of such scrap generated in the U.S. is four to five million pounds per year [Connor 2008].

Although there is not a large amount of post-consumer or end-of-life CFC waste in landfills or scrap yards today, the increasing implementation of CFCs in the aviation and automotive sectors and their subsequent decommissioning at their end-of-life means this issue must be addressed now. In September 2000, the European Union (EU) imposed an End of Life (EOL) Vehicle Directive (Directive 2000/53/EC) which regulates the disposal of vehicles. Requirements mandated in this directive included: 1) minimum 85% reuse/recover and 80% reuse/recycle of ELVs by January 2006; and 2) minimum 95% reuse/recover and 85% reuse/recycle of ELVs by January 2015.

The directive also set recycling quotas for non-metals, limited the allowable energy recovery, and mandated that vehicle manufacturers be responsible for the recovery cost [Bassam et al. 2011]. In 2005, a committee concluded that the reuse/recycle targets set in the 2000 EU Directive were high and unlikely to be achieved by 2015, and recommended keeping the 2006 target of 80%. The committee also recommended keeping the 2015 95% reuse/recover target and make the dismantling of non-metallic parts optional instead of mandatory [Bassam et al. 2011]. The European Parliament's Safety Division issued another study assessing the implementation of the 2000 Directive and found that Sweden and the Netherlands reached the 85% target, and Belgium met the 80% reuse/recycle target in 2005, but some countries had not fully met the directives. Factors hindering the implementation of these directives encompass a lack of economical technology to recover and recycle polymeric materials from shredder residue and the lack of profitable markets for the recycled polymers [Bassam et al. 2011]. Other factors include a lack of resources to implement the directives, particularly for vehicles already in service, and differing disposal regulations in different member countries [Bassam et al. 2011].

In Japan, an End-of-Life Vehicle Recycling Law went into effect January 1, 2005, mandating that all automakers charge car buyers with recycling fees; vehicle owners be responsible for taking their ELVs to an authorized recycling facility; manufacturers properly collect and dispose of fluids, airbags, and shredded residue; and the vehicle owners bear the cost for treating ELVs [Bassam et al. 2011].

The handling of ELVs in North America is not as progressive. In Canada, the Ontario Automotive Recyclers Association has a "Code of Practice" for its members, and Environment Quebec has a "Best Practice" guide for the ELV industry [Bassam et al. 2011]. In the U.S., regulatory activities have only occurred at the state level for the past couple decades, and have primarily dealt with labeling liquid containers, limits on dismantler storm water runoff, restrictions on mercury-containing components in landfills, and classifying shredder residue as hazardous waste in California. In 2005, automotive shredder residue was exempted from the waste generation fee in Ohio [Bassam et al. 2011].

Although there are currently no statutory recycling standards that U.S. automakers must meet, from a pragmatic viewpoint CF reclamation and recycling will be needed if CFCs are implemented in mainstream vehicles. Even at a modest level of 4.5–5.4 kg of CFC per mainstream vehicle, at current automotive scrap rates there would be about 45,359 tonnes of CFC scrap from passenger vehicles annually [Sullivan 2006]. Some of the leaders in the carbon fiber recycling industry credit the Aircraft Fleet Recycling Association (AFRA) for fostering scrap generator and CFC recycler cooperation and growing the recycling industry [McConnell 2011]. AFRA is a consortium of U.S. and European companies that was formed in the mid-2000s to combine their efforts working on the recovery of CF from manufacturing waste and end-of-life airplane scrap [Rush 2007; McConnell 2011].

6.2 Reclaiming and Recycling CFCs

There are a handful of methods for recovering clean carbon fibers from CFC waste: mechanical (milling/shredding), pyrolysis, fluidized bed oxidation, chemical, and supercritical fluid.

Milling/shredding/crushing or other mechanical means break down the CFC to small pieces that are separated via sieving into powdered products (rich in resin) and fibrous products (rich in fibers) [Pimenta and Pinho 2011]. The process reduces the fibers to very small lengths and does not recover individual fibers. This carbon fiber recyclate is of lower value than other forms of recyclate and is typically used for filler for injection molding of lightweight materials for non-structural parts.

Processes that recover the fibers from the CFC by breaking down the polymeric matrix (typically a thermoset) are called fiber reclamation processes, and include pyrolysis, fluidized bed oxidation, and chemical reclamation.

Pyrolysing CFCs breaks down the thermoset matrix resin through heating to 450–700°C in an inert (oxygen-free) environment. At this temperature the polymeric matrix is volatilized into lower-molecular weight molecules while the fibers remain inert. After pyrolysing, the mechanical properties of the recovered carbon fiber are about 80–90% of the virgin fiber properties. After pyrolysis, the carbon fiber recyclate can be milled or chopped. Milled recyclate is either pelletized or dry pressed, and then compounded for injection molding. Chopped recyclate, which can be ¼–1 in (6–25 mm) in length, can also be compounded for injection molding, or can be made into a non-woven for applications such as electromagnetic interference (EMI) shielding.

Pyrolysis research and development is also being done with microwave pyrolysis to raise the yield of reclaimed carbon fiber by eliminating char while having a shorter overall processing time and smaller scale equipment [McConnell 2011]. A continuous microwave approach has been implemented by Firebird Advanced Materials Inc. (North Carolina, U.S.) [Pimenta and Pinho 2011; McConnell 2011]. Work at Argonne National Laboratory (ANL) showed that a single-step pyrolytic process could recover very high yields of carbon fiber from thermoplastic or thermoset composites [Sullivan 2006]. Evaluation of these recovered fibers by Oak Ridge National Laboratory (ORNL) found the recovered fibers had similar diameter, density, morphology, and surface chemistry to those of virgin fibers. The ANL pyrolytic process is projected to be economically viable if the CF has a nominal value of $1.50 per pound [Sullivan 2006].

Fluidized bed oxidation is another thermal fiber reclamation process. It entails combusting the polymer resin in oxygen-rich flow at elevated temperature, such as air at 450–550°C. Scrap CFC fragments about 1 in (25 mm) in size are fed into a bed of silica on a wire mesh. Hot air streams through the bed and breaks down the resin. The oxidized resin molecules and the stripped fiber filaments are carried up with the air stream; a cyclone then extracts the fibers out of the air stream, and an afterburner fully oxidizes the resin. Any heavier metallic pieces that were in the scrap sink into the bed of silica. The University of Nottingham (UK) has been working with the fluidized bed process for years and has produced nonwoven mats of recycled carbon fiber (rCF) [McConnell 2011].

The other main type of fiber reclamation is through chemical means, based on a reactive medium at low temperature (typically <350°C). Chemical methods include catalytic solutions, benzyl alcohol, and supercritical fluids [Pimenta and Pinho 2011]. The process decomposes the resin into relatively large oligomers and leaves the CFs inert. Catalytic solution reclamation processes were developed over a decade ago, whereas benzyl alcohol and supercritical fluid (SCF) processes are much more recent approaches. The SCF has liquid-like density and dissolving power but gas-like viscosity and diffusivity, which enable the SCFs to penetrate porous solids and dissolve organic materials. At atmospheric conditions, the fluid is relatively benign [Pimenta and Pinho 2011]. Adherent Technologies Inc. (New Mexico, U.S.) uses a proprietary low-temperature liquid in its automated and continuously run chemical reclamation process to recycle manufacturing waste and end-of-life components. Its rCFs are either milled or chopped, and the resin products are recovered as fuel or chemical feedstock [Pimenta and Pinho 2011].In work by Knight et al. [2010], water was used as the supercritical fluid with KOH catalysts to reclaim CF from a single layer of woven Hexcel 8552/IM7 high-performance epoxy prepreg. The process produced a very high efficiency of resin elimination and clean fibers that had little change in tensile strength. This process could potentially recover woven fabric in its original format as opposed to random mat, leading to remanufactured composites from these rCFs that have better performance due to this better retention of fabric architecture and fiber properties [Knight et al. 2010].

A comprehensive overview of the state of the art and market outlook for CFC recycling operations is presented by Pimenta and Pinho [2011]. Each process has advantages and disadvantages: pyrolysis is the only process currently with commercial-scale implementation but requires a tremendous amount of energy; chemical methods are advantageous for maintaining mechanical properties but require the use of hazardous chemicals; and fluidized bed is attractive for end-of-life components and contaminated waste but causes more mechanical degradation [Pimenta and Pinho 2011]. Chemical processes reclaim the fiber and have the potential to recover resin products as fuels or chemical feedstock. The SCF process is quite new but causes very little mechanical degradation of the fibers and allows recovery of useful chemicals from the matrix [Pimenta and Pinho 2011].

Nguyen et al. [2007] investigated the use of microwave irradiation to degrade CFC: 12K CFC fabric was irradiated at 2450 MHz frequency at 400 or 700 W of energy. Exposure time at 400 W was varied from 20 seconds to 23 minutes, and exposure time at 700 W was varied from 5 seconds to 20 minutes. The effect of irradiation in terms of CFC degradation was evaluated by three-point bending modulus and strength, the latter of which would indicate delamination tendencies. Scanning electron microscopy (SEM) was used to assess crack distribution in the samples. At both power levels, the reduction in bending strength plateaued after approximately 20 minutes. The reduction in bending strength was approximately 33% at 400 W and 45% at 700 W. The elastic modulus did not change appreciably. This indicates that the microwave irradiation weakened the matrix by creating voids and/or cracks, and SEM analyses confirmed that the irradiation caused delamination at the interface of the fiber and matrix.

Researchers at North Carolina State University [Heil, Litzenberger, and Cuomo 2010] evaluated seven recycled carbon fibers from a third party for quality and appearance using SEM, X-ray photoelectron spectroscopy (XPS), and single filament tensile testing. SEM and XPS showed most recycled fibers had residual matrix material. Diameters of recycled fibers were not significantly different from virgin fibers, but pitting observed could explain the reduced tensile strength of single filaments. Intermediate modulus fibers showed about 55% strength retention, and standard modulus fibers showed about 80% strength retention. Tensile modulus of the recycled fibers was comparable to their virgin counterparts. In related work [Heil et al.2009] it was found that apparent fiber-matrix interfacial adhesion of intermediate modulus recycled fibers was as good as or better than that of virgin fiber, possibly from the observed recycled fiber surface morphology providing better mechanical interlocking. Conversely, the residue observed on the standard modulus recycled fibers could explain its reduced adhesion.

6.3 Implementation of Recycled CFCs

After carbon fibers are reclaimed by one of the available approaches, they need to be re-impregnated with a new matrix. Various options for remanufacturing rCFs include injection molding , bulk molding compound compression, compression molding of non-woven forms of rCFs, and sheet molding compounds.

Because there is extensive fiber breakage during the recycling process, and some additional breakage during remanufacturing, there is significant reduction in tensile properties of the rCF composite. However, in a study performed at Imperial College by Pimenta and Pinho [2011], it was found that the in-plane fracture toughness was increased by small amounts of unpyrolysed resin bonding fiber bundles together.

Several companies are intent on making recycled carbon fiber readily available through their carbon fiber composite recycling facilities set for commercial operations. Recycled Carbon Fibre Ltd. (West Midlands, UK; formerly Milled Carbon Ltd.) uses a continuous pyrolysis process that has the capacity to process 2000 tonnes of CF waste each year [McConnell 2011]. It also mills woven material to be used as fillers and cuts unidirectional fabrics to desired lengths [Heil, Litzenberger, and Cuomo 2010]. Materials Innovations Technologies, LLC (MIT-LLC) (Fletcher, NC, U.S.) has facilities to reclaim carbon fiber from manufacturing scrap and end-of-life parts using pyrolysis after a preliminary step of chopping the feedstock to a consistent length [Pimenta and Pinho 2011]. The recyclate is then processed into preforms for molding into parts. Pyrolysis processes are also established in industry in Japan, Germany, and Italy [Pimenta and Pinho 2011].

Because rCF is discontinuous and often tangled, there are difficulties using it in existing composite manufacturing processes. MIT-LLC is targeting a variety of forms of the recycled fiber for processing into a final product. Figure 6.1 shows images of the form of recycled fiber, a preform made using their 3-DEP™ process, and a finished molded part. The 3-DEP™ process is explained and the viability of making complex three-dimensional net-shape preforms using recycled fibers is demonstrated by Janney et al. [2009].

Aside from their innovative and Co-Dep processes, the reclaimed chopped fiber can be feedstock in thermoplastic compounding, used in LFT molding, or formed into Ultratek Mat (UM) of non-woven rolls of fiber for use in thermoset or thermoplastic molding.

Work by Heil and Cuomo [2011] presents the use of extrusion compounding and injection molding to produce thermoplastic recycled carbon fiber composites from thermoplastic manufacturing waste. They also used a wetlay process to make randomly oriented fiber mat preforms composed of virgin and recycled standard and intermediate modulus fibers.

Recycled carbon fiber   3-DEP™ pre-form   Molded part

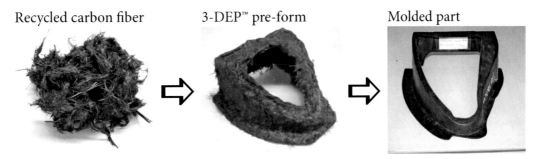

Figure 6.1. Reclaimed carbon fiber to compression molded part (Corvette lower wheelhouse support) by MIT-LLC.

(Images courtesy of MIT-LLC)

In a four-year UK-funded research program, Fibrecycle has made good progress toward developing low-cost, high performance CF materials from waste streams. Press molded carbon/PET composite laminates made from rCF had 90% of the tensile modulus and at least 50% of the tensile strength of an equivalent laminate made with virgin fibers [Project 2011]. Fibrecycle consists of six partners: Advanced Composites Group Ltd., Tilsatec, Sigmatex, Exel Corporation, NetComposites, and the University of Leeds.

6.4 Closing Thoughts

Worldwide, automotive companies are facing some challenging energy and environmental issues. In the U.S., the 2010 Corporate Average Fuel Economy (CAFE) regulation increasing fuel economy from 27 to 35 miles per gallon by 2016 has already resulted in concerted efforts to implement more lightweighting materials, including CFCs, in vehicle design. Means to recycle CFCs and other lightweighting materials must be developed to just maintain the current level of recyclability of vehicles made predominantly with steel. The CFC recycling industry is still in its infancy, and the processes are expensive and complicated. The industry has formidable requirements, including consistent scrap availability, appropriate size reduction technologies, established process parameters, the infrastructure for material collection, and standardization of recyclate properties [McConnell 2011]. The technical and economic issues with recycling/reusing CFCs are best developed during vehicle design to aid in both the recovery of the material as well as potential implementation of the recyclate back into a vehicle.

References

Bassam, Jody, Joseph A. Pomykala, J. Spangenberger, and Edward J. Daniels. 2011. "Recycling of the Changing Automobile and Its Impact on Sustainability." SAE Paper No. 2011-01-0853. doi: 10.4271/2011-01-0853. SAE International, Warrendale, PA. 2011.

Connor, Myles Linden. 2008. "Characterization of Recycled Carbon Fibers and Their Formation of Composites Using Injection Molding." A thesis submitted to the Graduate Faculty of North Carolina State University In partial fulfillment of the Requirements for the degree of Master of Science, Materials Science and Engineering.

Heil, Joseph P., and Jerome J. Cuomo. 2011. "Recycled Carbon Fiber Composites Using Injection Molding and Resin Transfer Molding." *16th International Conference on Composites Structures* no. ICCS 16.

Heil, Joseph P., Davis R. Litzenberger, and Jerome J. Cuomo. 2010. "A Comparison of Chemical, Morphological, and Mechanical Properties of Carbon Fibers Recovered from Commercial Recycling Facilities." SAMPE 2010 - Seattle, WA. May 17-20, 2010.

Heil, Joseph P., Michael J. Hall, Davis R. Litzenberger, Raphael Clearfield, Jerome J. Cuomo, Pete E. George, and William L. Carberry. 2009. "A Comparison OF Chemical, Morphological, AND Mechanical Properties OF Various Recycled Carbon Fibers." SAMPE 2009 - Baltimore, MD. May 18-21, 2009.

Janney, Mark A., W. Leroy Newell, Ervin Geiger, Neal Baitcher, and Tod Gunder. 2009. "Manufacturing Complex Geometry Composites with Recycled Carbon Fiber." SAMPE 2009 - Baltimore, MD. May 18-21, 2009.

Jody, B. J. Junior, Joseph A. Pomykala, Jeffrey S. Spangenberger, and E. J. Daniels. 2009. "Impact of Recycling Automotive Lightweighting Materials on Sustainability." SAE Paper No. 2009-01-0317. SAE International, Warrendale, PA. 2009.

Knight, Chase C., Changchun Zeng, Chuck Zhang, and Ben Wang. 2010. "Recycling Carbon Fiber Composites Using Supercritical Fluids." SAMPE 2010 - Seattle, WA. May 17-20, 2010.

McConnell, Vicki P. 2011. "Reinforced Plastics - Launching the carbon fibre recycling industry." *Reinforced Plastics.com.* March 29, 2010. (http://www.reinforcedplastics.com/view/8116/launching-the-carbon-fibre-recycling-industry.com Accessed July 29, 2011).

Nguyen, Phuong Ngoc Diem, Susan A. Roces, Florinda T. Bacani, Masatoshi Kubouchi, Sakai Tetsuya, and Piyachat Yimsiri. 2007. "Degradation Behavior of Carbon Fiber Reinforced Plastic [CFRP] in Microwave " Asean Journal of Chemical Engineering (Vol. 7, No. 2):157-164.

Pimenta, S. and S. T. Pinho. 2011. "Recycling carbon fibre reinforced polymers for structural applications: technology review and market outlook." *Waste Management* no. 31 (2):378-92. doi: 10.1016/j.wasman.2010.09.019.

"Project develops low cost blended carbon fibre yarns and fabrics." 2011. JEC Composites no. May 17, 2011 (http://www.jeccomposites.com/news/composites-news/project-develops-low-cost-blended-carbon-fibre-yarns-and-fabrics Accessed August 14, 2011).

Rush, Susan. 2007. "Carbon fiber: Life Beyond the Landfill." *High Performance Composites* (http://www.compositesworld.com/articles/carbon-fiber-life-beyond-the-landfill Accessed August 22, 2011).

Sullivan, Rogelio, A. 2006. "Automotive Carbon Fiber: Opportunities and Challenges." *JOM* no. November 2006.

Chapter Seven

Implementation and Longevity

For a successful technology, reality must take precedence over public relations, for Nature cannot be fooled.
—Richard P. Feynman

7.1 Design and Modeling

As with any material, but perhaps more so with CFCs, components should be designed for manufacture, performance, and service, as applicable. Digital tools play a huge role in processing the raw materials into the constituents of CFCs, from process control to chemical modeling the kinetics of chemical reactions. They are also important in manufacturing of composites, supporting processes ranging from tool machining to automated manufacturing steps, such as kit-cut design, automated tape laying, and braiding machines, to component trimming and drilling. Design and modeling tasks are aided by digital tools as well.

Digital tools are employed by designers, analysts, and manufacturers. The knowledge held within each of these disciplines is interdependent, and individuals working in one discipline often are not trained or are not required to perform the functions in the other disciplines [Collier 2011]. Typically, a design department designs the composite component and hands off the design to the structural analysts to determine the necessary fiber/resin construction for it to carry the design loads. The load-based design is then adjusted by the designers and manufacturers based on the available manufacturing processes and their limitations. Information and data are handed off, but there is no real-time design iteration among these disciplines, resulting in designs that can be 15–20% overweight if optimization tools are not utilized [Collier 2011]. Many software tools that have vastly improved composite design, analysis, and manufacturing are becoming better interconnected across these disciplines and provide seamless data exchange between computer aided design, finite element analysis, and manufacturing simulation.

The following sections briefly discuss digital tools as applied to fiber orientation, computer aided engineering, and mechanical performance modeling. Digital tools associated with process modeling, which includes molding (injection, compression, RTM, and its variants), fiber placement, forming, and joining are not discussed aside from the fiber orientation subset because of its importance in properly modeling composite performance.

7.1.1 Fiber Orientation and Composite Construction

Carbon fibers are used for their high tensile modulus and strength, and hence it is important in designing a component to have the carbon fibers loaded predominantly in the tensile direction. For components subjected to multiaxial loading, a uniaxial orientation of the fibers will be much less effective than fibers oriented in multiple directions or randomly. Software tools exist to help the designer construct the composite to optimize the use of carbon fibers to carry the component's design loads.

There are software tools that model the fiber orientation of many of the fiber architectures discussed in Chapter 3. Some model fiber size, orientation, and distribution of discontinuous fiber-reinforced injection molded components, while others model the draping of fabrics around complex geometries. Most commercially available software for modeling fiber orientation was originally developed for a particular composite construction or manufacturing process, and many have expanded their capabilities to other constructions or processes as demand warranted. Modeling drapage is necessary to provide accurate fiber orientation of tape or fabric material as it drapes around complex geometries. Wrinkles, folds, or stretches in the fabric can detrimentally affect the end composite both mechanically and aesthetically.

7.1.2 Computer Aided Engineering

Computer Aided Engineering (CAE) is used throughout the whole process of fabricating CFC components, from tool and component design to modeling the manufacturing process(es) and finally modeling the end-part behavior.

CAE packages exist that develop finite element meshes of a component, considering part geometry and layup/architecture, for subsequent analysis. Other CAE packages exist for simulating fabrication processes, conducting structural analyses, or analyzing fatigue performance. CAE has aided structural optimization for decades, with the digital age permitting systematic numerical optimization to structural design. Advances in computing power and better integration between disciplines have enabled topology optimization to take root in several industries, including automotive. Topology optimization uses finite element techniques with mathematical optimization methods to optimize the layout of material within a defined space to meet performance targets under defined loads and boundary conditions, and even manufacturing constraints. A-E-S Europe GmbH has described topology optimization as a finite element analysis algorithm that evolves the optimal lightweight shape for a structural design, similar to what nature does in bones, trees, and bird wings [Gardiner 2011]. Mass is redistributed throughout the structure to produce a more homogeneous stress distribution, and as a result overall mass is usually reduced. Topology optimization is becoming necessary in the aerospace industry to successfully optimize their large and complex composite structures [Gardiner 2011], and the automotive industry can well benefit from their developments.

7.1.3 Mechanical Behavior of CFCs

Mechanical properties of the composite that are important in component design include stiffness, strength, thermal expansion, electrical conductivity, thermal conductivity, permeability, acoustic emissivity, energy dissipation, strain to failure, and fatigue.

Wider use of CFCs in vehicle structural components depends in large part on understanding and controlling their deformation under the impact conditions. Structural components surrounding the passenger compartment must dissipate impact energy in a controlled and stable fashion. In the current vehicles based on unibody steel design, the impact energy is dissipated by large plastic deformations brought about by deep bending and folding of the vehicle's structural elements. The macro-scale deformation and the underlying recurring kinematic micro-mechanisms are reasonably well understood and predictable, and supported by an extensive empirical experience on crashworthiness design with metals. The crash of conventional steel-based vehicles can be modeled within the framework of large deformation/large displacement continuum mechanics and the finite element method (FEM). Cracks can occur when large deformations have accumulated, and if needed, fracture mechanics is available to model crack propagation in ductile materials.

The situation is quite different for crashworthiness of fiber reinforced composite materials, and in particular, for CFCs. Here cracks are the rule, not the exception. The CFCs have comparably low ductility and exhibit strong anisotropy of mechanical properties. Their mechanical response is governed by a multitude of deformation mechanisms in their constituents and interfaces. The main dissipative mechanisms are the formation of micro and macro cracks, progressive fracturing, and failures operating on many length scales. These processes are inherently difficult to quantify via experiments and equally difficult to describe in the framework of continuum mechanics and FEM. Although there has been a tremendous amount of modeling work done on CFCs in the racing arenas such as Formula 1 [Savage 2008], many assumptions and estimates still need to be made when modeling CFC structures, and mechanical testing on representative coupons and components continues to be done in parallel with computer analysis. For vehicle designers to be able to use the CFCs to their full capacity, they need to have in-depth understanding of the governing deformation mechanisms and be able to model them using practical computational simulation tools. Design and modeling for vehicle crashworthiness have some unique requirements. Whereas the scope of standard composite failure analyses in aerospace designs end at the indication of imminent onset of failure and determination of the limit load of the material, the vehicle crashworthiness modeling must consider the entire loading event the material goes through during progressive crush.

Typical crashworthiness simulation for CFCs is performed using the FEM explicit time integration code with a combination of multiple submodels that attempt to account for the main mechanisms of underlying deformation. Taking a laminate composite as an example, a model can include a multi-layered shell and/or solid finite elements, constitutive models for each of the lamina, material property degradation models, failure criteria for the laminas and laminate, lamina and laminate removal models, delamination models, and crash front models. Each of the submodels describes mechanical phenomena that operate on its own length scale or range of scales and may interact with other submodels and phenomena. The above list is not all-inclusive and very often is reduced or expanded based on the complexity of the problem and the ability to consolidate multiple submodels into effective continuum formulations.

Major groups of submodels include homogenization formulations, FEM element formulations, constitutive models for specific composite types, failure criteria/failure surfaces, treatment of localization and strain softening, delamination models, and crush progression models.

To date, most packages are capable of simulating and oftentimes predicting the elastic behavior of CFCs, using fundamental properties of the constituent materials and their construction, with accuracy that is likely adequate for the automotive industry. However, the major issue that has yet to be overcome is the ability to predict behavior beyond the onset of plastic deformation.

The Department of Energy recognized this hurdle facing the industry and in late 2011 awarded funding to two programs — one led by Plasan Carbon Composites, and one led by the U.S. Automotive Materials Partnership (USAMP) — to validate existing models for predicting the behavior of CFCs in automotive crash applications. The outcomes from these two programs are expected to have a significant positive effect on the use of CFCs in automotive structural applications.

7.2 Physical Testing

Although the spectrum of mechanical and physical properties is important for appropriate design and use of CFCs, the ones of most concern to automotive engineers and managers who are considering implementing CFCs are the mechanical stiffness, strength, strain to failure, and fatigue performance — in the various loading modes and environmental conditions that apply.

In 2002, the Automotive Composites Consortium (ACC) within USAMP reported on a multi-year program to develop a design and manufacturing strategy for a composite-intensive body-in-white (BIW). A complete dynamic crashworthiness analysis on the BIW was not part of the project because the requisite tools for predicting the crashworthiness of composite structures were not yet available. To accomplish such an endeavor would have required extensive empirical data on both components and full vehicles [Boeman and Johnson 2002]. Although much work has been done since then, the needle has not moved far when it comes to obtaining experimental data on the spectrum of CFCs applicable to the automotive industry.

Probably the main impediment to progress in this area is that, compared to metals, fiber reinforced composites present additional testing issues, particularly in the two areas of concern for automotive structural components — high strain rate testing to understand crash behavior, and fatigue testing to understand long term durability. Adding to the problem is the splendid variety of CFCs which, due to their different constituent makeup or architecture, all need evaluation.

The small specimen length needed to attain high strain rates within the specimen gauge is in conflict with the size of specimen needed to represent the bulk material; often the gauge length and cross-sectional area of current high-rate specimens is of the same scale as the fiber length or fiber architecture. A larger specimen would overcome these problems but introduce new issues such as exceeding equipment capacity, introducing the load properly into the material, achieving dynamic equilibrium within the time frame of the test event, and natural oscillatory vibrations within the specimen.

To achieve balance between the competing small and large specimen size drivers, a specimen geometry and size can be designed for a particular composite system and fiber architecture. Although this is not practical or appropriate for comparing different composite systems, this is the point at which the industry is currently.

The sample size issue also exists in fatigue testing of composites, along with other issues. Because of hysteretic heating, fiber reinforced plastics, and particularly thermoplastics, need to be tested at frequencies typically at least an order of magnitude slower than metals. Test frequencies are typically 2–5 Hz, making the time to reach run-out (10^6 or 10^7 cycles) extremely long.

CFC components are also difficult to test dynamically to failure to understand crash behavior. Their high stiffness, high strength, and lack of ductility present challenges for the testing equipment to have the load capacity along with the ability to ensure proper alignment and load introduction, as well as the ability to reach a speed that represents strain rates experienced in automotive crashes. There is also the difficulty of measuring the progression of failure throughout the event because the brittle nature causes it to be so abrupt, and because the nonhomogeneous and nonisotropic material makes it difficult to know where the onset of failure will occur.

A fair amount of high rate and fatigue testing has been done on fiber reinforced plastics. Excellent starting points for the interested reader are the works by S. Hill at the University of Dayton Research Institute (UDRI) for high strain rate behavior of composites, and the work by Sandia National Laboratories (SNL) for fatigue of composites.

7.3 Quality Control

Quality control has become paramount to every component in the automotive industry, similar to aerospace but not quite to the same degree of statistical certainty. Aside from being mandatory, quality control is the means to assure that the component is composed of the right materials, processed under the proper conditions, fabricated into a component under the right conditions, the culmination of which has been verified to provide a certain type and level of performance. The automotive industry has well-established quality control systems in place for its traditional metallic components, and can look to the aerospace industry for guidance in areas specific to CFCs and their assembly to dissimilar materials. As Cohen stated in his keynote address at SAMPE Europe in 2010 [Cohen 2010], a couple key issues pertaining to quality control of CFCs are what technologies are currently being used, and what new technologies should be developed.

There are several stages in the product development process to which industry-accepted or standardized practices can be applied. Starting with raw material supplies of carbon fiber and resin, or pregreg, is the certification of material content and quality provided by the suppliers. Next is the processing of the constituents into a CFC, which involves quantities of material, processing conditions including temperatures and pressures and

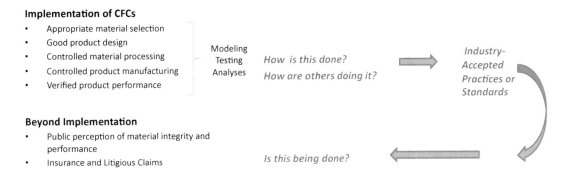

Implementation of CFCs
- Appropriate material selection
- Good product design
- Controlled material processing
- Controlled product manufacturing
- Verified product performance

Modeling
Testing
Analyses

How is this done?
How are others doing it?

Industry-Accepted Practices or Standards

Beyond Implementation
- Public perception of material integrity and performance
- Insurance and Litigious Claims

Is this being done?

Figure 7.1. The need for industry-accepted or standardized practices for quality control and beyond.

their rates of change, layup of tapes or fabrics, fiber size/orientation/distribution, and any modeling and analyses of the processing. Next is the manufacture of the part/component, which has some aspects similar to processing as well as thicknesses, trimming/drilling/attachments, surface quality for appearance, fatigue or high rate performance, and any modeling and analyses of the manufacturing. Last is the performance, at the component level and at the vehicle level, which involves testing, modeling, and analyses of behavior under the appropriate loading, such as flexural, tensile, creep, fatigue, high strain rate, hot and cold temperature, and environmental exposure.

In each stage, it is essential that industry-accepted/standardized practices be demonstrated in terms of quality control, document retention, test records, and internal test performance criteria versus regulated performance criteria (such as Federal Motor Vehicle Safety Standards, FMVSS). CFCs are complex material systems whose performance directly depends on their constituent materials, processing, and manufacturing conditions. Having quality control documents proving how Company A assures the right parameters have been controlled and checked is mandatory; showing that they do it the same way that Companies B, C, and D do it — i.e., using industry-accepted or standardized practices — is essential to instilling consumer confidence in the material, the product performance, and the company.

In our current-day society, the automotive industry is fraught with insurance claims and lawsuits. Having adequate documentation of the quality controls exercised throughout the entire product development process is the single most helpful aid in determining if in fact there is an issue with the product and/or material, and if not, in providing defense against wrongful claims. Figure 7.1 highlights these key aspects of quality control.

7.4 Non-Destructive Evaluation

Non-destructive evaluation (NDE) techniques are used to observe manufacturing quality conditions and to detect in-service damage. The anisotropic and nonhomogeneous nature of CFCs makes damage detection much more challenging than in quasi-isotropic materials such as metals. Additionally, the damage mechanisms in CFCs are different than in metals, and include fiber/matrix debonding and delamination. Most sensing techniques that work satisfactorily for metals are not reliable for sensing the behavior of composites [Singh et al. 2009]. Radiography, acoustics, ultrasonics, and eddy current techniques have been extensively studied and used [Warraich et al. 2009]. Radiography was replaced by ultrasonics due to safety hazards and other limitations, and conventional ultrasonics have in turn demonstrated their own limitations, such as the detection of a defect being proportional to a predefined threshold level and the method being vulnerable to noise [Warraich et al. 2009]. To date, standard NDE methods have limited ability to observe manufacturing quality variations and in-service damage evolution because most methods provide only an indication of the presence of damage and little information about the defect or damage characteristics [Washabaugh, Grundy, and Goldfine 2010].

Digital signal processing techniques have aided in removing noise from ultrasonic signals and in localizing defects, but they sometimes result in critical information being lost if the filtered frequencies carry the topology of the defect [Warraich et al. 2009]. Kalman filters successfully reduce signal noise but limit the possibility of detecting multiple defects [Warraich et al. 2009].

Warraich et al. stochastically estimated the location of defects using probability functions considering a range of possibilities and assumptions, preventing loss of information that is carried by the ultrasonic A-signal. Using CFCs with artificial delamination (via Teflon film between plies) and porosity (via microsphere dispersion), their technique successfully estimated the location of the induced defects over a range of specimen thicknesses, and successfully detected multiple defects in an ultrasonic signal. Their method permits stochastic localization of defects in one dimension using ultrasonic A-scan signals, and in two dimensions using C-scan images [Warraich et al. 2009].

When applied to detecting in-service damage, NDE is often referred to as structural health monitoring (SHM). SHM has become fairly common for aerospace structures during production and in-service and for some civil structures (e.g., bridges) while in-service. SHM is a process to detect damage and thereby prevent catastrophic failure by either predicting the onset of failure or implementing repair. Damage is manifested as a change in mechanical behavior of the material, which can affect the component's performance at some point in its design life. For materials that exhibit brittle behavior, the lack of ductility translates to lack of warning.

SHM can be separated into those that detect in-service or real-time and those that detect during an isolated inspection. Those that can detect strain in composite structures in real-time include optical fibers, electrical resistance strain gauges, acoustic emissions, and piezoelectric transducers [Singh et al. 2009]. All devices that can detect in real time are either surface-mounted or embedded in the structure, which requires that they withstand the composite processing and assembly; be compatible with the composite chemistry; be able to withstand the same mechanical, chemical, and environmental loads as the composite; and enable wireless communication. Singh et al. presented their work on using inherently conducting polymers (ICPs) as strain sensors. Although the resistance measurements of the ICP film on carbon/epoxy composite did not correlate as well with mechanical strain as they did on Teflon substrates, the researchers postulated that the inherent conductivity of CFC caused noise in the signal and that the ICP technology shows promise [Singh et al. 2009].

Washabaugh, Grundy, and Goldfine [2010] studied the use of magnetic field-based eddy current sensors as an NDE method for detecting impact damage or monitoring stress in graphite fiber composite materials. A micromechanical model was developed in conjunction with the study to gain insight into the constituent properties being monitored. The result is a model-based NDE method that could detect some damage, such as delamination, directly from NDE, and in other situations the model enabled NDE information to be correlated to the material's mechanical state of stress. This model-based approach could be applied to other NDE methods such as dielectrometry, ultrasonics, and thermography [Washabaugh, Grundy, and Goldfine 2010].

Although detecting damage during an out-of-service or manufacturing inspection avoids some of the technical challenges of real-time monitoring, it can still be a formidable task. The inspection technique should be capable of detecting potential manufacturing issues such as delamination in laminate constructions, as well as porosity and incomplete fiber wetout, and potential damage indicators such as microcracks, fiber/matrix debonding, and fiber/tow pullout. Ultrasonic systems can operate in different modes: pulse-echo, resonance, through-transmission, and eddy current. Through-transmission is often required to penetrate parts. Systems also need to have a database that correlates defect level and attenuation value; the physical characterization of a defect level can be established by sectioning samples and using conventional metallographic techniques of mounting, polishing, and microphotographing [Engelbart 1999].

Krajca and Ramulu [2001] present a sample ultrasonic C-scan image of pierced holes and the corresponding micrograph of one of the holes, revealing partial through-thickness delamination (Fig. 7.2).

In April 2010, MISTRAS Group Inc. joined forces with HITCO Carbon Composites Inc. to produce a 12-axis ultrasonic C-scan imaging gantry scanning system for aerospace CFCs. It will have through-transmission and pulse-echo scanning capabilities [MISTRAS 2010].

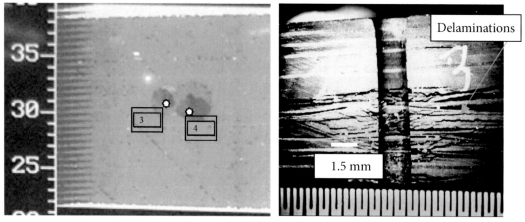

Figure 7.2. Ultrasonic C-scan image revealing partial through-thickness delamination [Krajca and Ramulu 2001].
(Used with permission of SAMPE)

Ultrasonic techniques are currently the most commonly used, but other flaw detection techniques in use and under further development include thermography, shearography, and laser vibrometry. For example, Materials Sciences Corporation (MSC) developed an in-line scanning thermography system with Physical Acoustics Corporation (PAC) under the Small Business Innovation Research (SBIR) program.

7.5 Repair

The use of CFCs in mainstream vehicles will necessitate the development of repair techniques in lieu of 100% replacement when parts are damaged. Because insurance companies bear the costs of repair/replacement for damaged, insured vehicles, that industry has a vested interest in having accurate and field-ready damage detection techniques to determine the extent and severity of damage, and sound repair techniques cascaded throughout the market. Damage caused by even minor accidents can range from cosmetic to structural. The traditional steel and aluminum metals used in today's vehicles deform plastically under impact and can be "hammered" back into shape if the deformation is not too excessive. Auto repair shops can "hammer" bumper beams and front rails, for example, back into shape using industry-accepted practices endorsed by the insurance industry. However, the brittle nature of CFCs (and glass fiber composites) means they undergo no appreciable plastic deformation before fracturing. Without proper repair techniques, a cracked CFC component would have to be completely replaced. The same is true for many composite materials used in aviation. Aluminum and steel are quite damage-tolerant materials, able to be dented or punctured and still hold together. When repair is necessary, it can be done at almost any auto repair shop. CFCs are not so damage tolerant and require almost immediate repair or replacement.

For the aerospace industry, SAE AMS3970 [SAE 2011] covers 120°C (250°F) vacuum curing repair prepreg and SAE AMS2980A [SAE 2006] covers the wet lay-up repair materials for carbon fiber/epoxy structures conforming to the requirements of the aerospace OEMs. (The actual repair procedures are not part of these specifications.) Because of the challenge of repairing composites having a thermoset matrix, which by definition cannot be melted and reformed, thermoplastics have been looked at to replace thermosets as the matrix material for some CFC applications where repair is foreseeable. In addition to being able to be softened or melted and reformed, thermoplastics typically have higher toughness and ductility than thermosets, making them less susceptible to cracking under light impacts. As discussed in previous chapters, thermoplastic composites can be cheaper and easier to produce, but have disadvantages too, such as poor dimensional stability and lower operating temperature.

The solutions for repair of CFCs in automotive applications will require input from the entire value stream including raw material suppliers, processors, composite fabricators, OEMs, the insurance industry, and repair shops. This topic was discussed at an automotive workshop on lightweight materials sponsored by Oak Ridge National Laboratory (ORNL) and Pacific Northwest National Laboratory (PNNL) in 2011, and participants anticipated the need for a cross-industry effort to establish accurate damage detection techniques, standardized repair techniques, certified CFC repair shops, and certified facilities to train and educate those in the field who will make the decision to repair or replace and who will execute the repair.

7.6 Closing Thoughts

Implementation and longevity of CFC components in mainstream vehicles hinge on a multitude of technical issues, covering raw materials, fabrication, assembly to other (CFC or non-CFC) components, and in-vehicle performance. However, addressing all the technical issues will not guarantee first-use or long lasting use of CFCs in mainstream vehicles. Acceptance of the material is also key — the acceptance of CFCs by OEMs through the inclusion of CFCs in their portfolio of materials from which they can design mainstream vehicles, and acceptance by the consumers with regard to cost and performance throughout the vehicle life, which inevitably includes damage and repair.

References

Boeman, Raymond G. and Nancy L. Johnson. 2002. "Development of a Cost Competitive, Composite Intensive, Body-in-White." SAE Paper No. 2002-01-1905. SAE International, Warrendale, PA. 2002.

Cohen, Leslie. 2010. "The digital thread: Transformation to automation." High Performance Composites. July 1, 2010 (http://www.compositesworld.com/columns/the-digital-thread-transformation-to-automation Accessed August 12, 2011).

Collier, Craig. 2011. "Purging conservatism from composite designs." *High Performance Composites.* September 2011. no. 19 (5).

Engelbart, Roger W. 1999. "On-Line NDE of a Commercial Transportation Beam Fabricated by CRTMTM." 44th International SAMPE Symposium, May 23-27,1999.

Gardiner, Ginger. 2011. "Topology optimization." *High-Performance Composites.* September 2011. no. 19 (5).

Krajca, Scott E. and M. Ramulu. 2001. "Abrasive Waterjet Piercing of Holes in Carbon Fiber Reinforced Plastic Laminate." 33rd International SAMPE Technical Conference - Seattle, WA - November 5 - 8, 2001.

"MISTRAS, HITCO develop large ultrasonic C-scan system. " 2010. *Composites World.* April 18, 2010. (http://www.compositesworld.com/news/mistras-hitco-develop-large-ultrasonic-c-scan-system Accessed August 12, 2011).

SAE. 2006. "AEROSPACE MATERIAL SPECIFICATION AMS 2980A Carbon Fiber Fabric and Epoxy Resin Wet Lay-Up Repair Material." SAE International, Warrendale, PA.

———. 2011. "AEROSPACE MATERIAL SPECIFICATION AMS3970 Rev. A (R) Carbon Fiber Fabric Repair Prepreg, 120 °C (250 °F) Vacuum Curing." SAE International, Warrendale, PA.

Savage, G. 2008. "Composite Materials Technology in Formula 1 Motor Racing."

Singh, Abhishek K., Dongsik Kim, Huaxiang Yang, Brady W. Pitts, Gregory J. Tregre, and Patrick J. Kinlen. 2009. "Structural Health Monitoring of Carbon Fiber Composites Using Inherently Conducting Polymeric Films." 41st International SAMPE Technical Conference - Witchita, KA - Oct 19-22, 2009.

Warraich, Daud S., Donald W. Kelly, Tomonari Furukawa, and Israel Herszberg. 2009. "Ultrasonic Stochastic Localization of Hidden Defects in Composite Materials." SAMPE 2009 - Baltimore, MD - May 18-21, 2009.

Washabaugh, Andrew, David Grundy, and Neil Goldfine. 2010. "Composite NDE Using Quasistatic Electromagnetic Methods." SAMPE 2010 - Seattle, WA May 17-20, 2010.

Chapter Eight

Concluding Thoughts

Obstacles are those frightful things you see when you take your eyes off your goal.
—Henry Ford

8.1 Manufacturing and Assembly in Legacy Plants

For decades, various forms of glass fiber reinforced plastics (GFRPs) have been used in mainstream automotive, and carbon fiber reinforced plastics (CFRPs or CFCs) have been used in the aerospace industry. However, mainstream automotive has been elusive to CFCs, with the major obstacles for years being cost and cycle time. Those obstacles are shrinking as a result of efforts made by researchers, material suppliers, processors, manufacturers, and OEMs over the past few years. As an industry, we can now see past the cost and cycle-time hurdles to face the challenge of how CFCs can be accommodated by OEMs in their legacy manufacturing and assembly plants.

Mass production of passenger vehicles was developed and remains centered on metals, whether it be part fabrication, joining and assembly, painting, or repair. CFC parts do not seamlessly slide into an OEM's manufacturing or assembly process. As expressed at an automotive composites conference not too long ago (2007), "true integration of composites into automotive design requires not the part-by-part nibbling of the last 20 years, but a wholesale rethinking of car design, starting from the ground up" [Sloan 2011, 2007]. Existing automotive plants have significant investments in equipment that are devoted to metal, with regard to how — and how quickly — metal materials are manufactured and assembled. As of 2009, after major restructuring in the mid-2000s, the "Domestic 3" (GM, Chrysler, and Ford) had 26 assembly plants in the U.S. (11, 9, and 6, respectively), 6 in Canada (2 each), and 8 in Mexico (4, 2, and 2, respectively). The foreign implants (Europe and Asia) have 18 plants in the U.S., 3 plants in Canada, and 5 plants in Mexico [Canis and Yacobucci 2010]. On average, the 32 "Domestic 3" assembly plants in the U.S. and Canada are decades older than the 21 plants of the foreign implants, as the foreign producers started making vehicles in the U.S. in the early 1980s, and most of their plants opened after 1991 [Klier and McMillen 2008]. It stands to reason that plants built or upgraded before 1991 were not envisioned to be working with carbon fiber composites for the high volumes of vehicles they output.

A comparable investment — and challenge — exists in the automotive workforce, from plant worker to designer to engineer, that is predominantly educated and trained to work with the traditional metal materials. According to the Congressional Research Service report "The U.S. Automotive Industry: National and State Trends in Manufacturing Employment" [Platzer and Harrison 2009], the number of workers in the U.S. motor vehicle manufacturing industry first fell below one million in 2007. The numbers have continued to drop over the past few years, but it is still a very large workforce that is predominantly much more familiar with metals than composites. And it is not just the OEMs that are lacking familiarity with composites. In a survey conducted by VISTAGY, it was revealed that only 56% of the composite design companies surveyed considered themselves knowledgeable in composites manufacturing and applied that knowledge during design [Survey 2011]. A similar issue of lack of experience and confidence in CFC design data and analytical tools was also identified by Sullivan [2006] as one of several barriers that must be overcome before CFCs can be widely implemented in the automotive industry.

At that 2007 automotive composites conference, it appeared that the OEMs were not sufficiently motivated to undertake the wholesale rethinking of car design that was indicated to be needed. Rather, it was suggested that a consortium from the composites industry should do this rethinking and redesigning of the automobile for composites [Sloan 2007]. Although such a consortium has not been formed, in the few years since then, there have been many agreements, partnerships, and joint ventures formed between OEMs and materials suppliers and/or equipment suppliers in the carbon fiber industry (Chapter 2). These strategic ventures between OEMs and suppliers are decisively moving carbon fiber composites into the automotive market beyond niche vehicles. An excellent example is BMW's Mega City Vehicle (MCV), which is to launch mid-2013. The MCV will contain CFC body parts produced in Landshut, Germany. With that goal in mind, BMW built a complete carbon fiber laboratory in Landshut in 2000, formed a joint venture with SGL Carbon for CFC expertise, and partnered with Schuler SMC to develop a new process for RTM and supply the CFC component production lines catering to the MCV production volumes [Automotive 2011].

It took a huge commitment by government and industry, really around the world (U.S., Japan, U.K.), to bring CFCs into aerospace. Although much technical information can be cascaded from aerospace to automotive, automotive performance and manufacturing requirements are significantly different, and aerospace materials and processes cannot simply be adopted by the automotive industry. Much work still must be done to get CFCs integrated into the mass vehicle market, and this will depend on the entire community — OEMs, automotive suppliers, the carbon fiber industry, resins suppliers, and equipment and tool manufacturers — to get it done.

8.2 Advancing with the Advancements of Other Materials

The portfolio of CFC material systems that is, or soon will be, available for use in large-volume vehicle production are amenable to a large number of different types of automotive components. However, not every component that could be made from CFCs will be made from CFCs, nor should they. In an ideal world, CFCs will be used where their mechanical and physical properties are best suited. In the real world, several other factors play in the material selection process, including but not limited to raw material cost, manufacturing cost, part-to-part cycle time, assembly to other vehicle components, capital investment, and life cycle (sustainability/recyclability). The material selected for a component is based on the performance and business case of that material compared to competing materials.

As reported by the National Renewable Energy Laboratory [NREL 2010], one of the goals of the Materials Technologies subprogram of the U.S. Department of Energy's

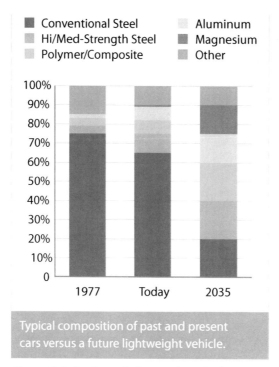

Figure 8.1. Projected change in vehicle
materials [NREL 2010].
(Courtesy of U.S. Department of Energy)

Vehicle Technologies Program is to demonstrate a cost-effective 50% weight reduction in passenger-vehicle body and chassis systems by 2015. Figure 8.1 illustrates the projected change in material composition to achieve that goal.

In a survey of engineers [Gehm 2011], 95% of respondents think materials are very important/important to a product's success. In 2011, 75% of respondents expect advanced composites to gain the most ground in vehicle design/engineering in the next decade.

Depending on the performance requirements, materials competing against CFCs can be GFRPs, aluminum, magnesium, and advanced high-strength steels. The industries surrounding each of these competing materials have been and continue to develop improved performance with lower cost or faster cycle time or fewer parts. Candidate materials for mass reduction are shown in Table 8.1 along with their relative cost to steel baseline [Warren et al. 2008].

Magnesium (Mg) has the lowest density of all structural metals, giving it the potential to reduce component weight by more than 60% [Report 2011]. Research at Pacific Northwest National Laboratory (PNNL) and Oak Ridge National Laboratory (ORNL) has characterized the behavior of Mg under high strain rates (relevant to automotive crash; see Fig. 8.2), demonstrated high-rate warm-forming of Mg sheet, and produced

Table 8.1. Cost-Comparison of Candidate Lightweighting Materials
[*Warren et al. 2008]* (Courtesy of U.S. Department of Energy)

Lightweight Material	Material Replaced	Mass Reduction (%)	Cost (per part) Relative to Steel
High Strength Steel	Mild Steel	10	1
Aluminum (Al)	Steel, Cast Iron	40 - 60	1.3 - 2
Magnesium	Steel or Cast Iron	60 - 75	1.5 - 2.5
Magnesium	Aluminum	25 - 35	1 - 1.5
Glass FRP Composites	Steel	25 - 35	1 - 1.5
Graphite FRP Composites	Steel	50 - 60	2 - 10+
Al matrix Composites	Steel or Cast Iron	50 - 65	1.5 - 3+
Titanium	Alloy Steel	40 - 55	1.5 - 10+
Stainless Steel	Carbon Steel	20 - 45	1.2 - 1.7

Figure 8.2. Crushed magnesium column.
(Courtesy of U.S. Department of Energy)

a preliminary design for a Mg-intensive vehicle front end that is less than 56% of the steel baseline. Issues with Mg components in vehicle design include joining, corrosion, repair, and its low ductility.

Aluminum (Al) is in the middle of the structural metals spectrum and is already well developed and implemented in automotive design [Report 2011]. Aluminum applications include body panels, body-in-white (BIW) structures, and powertrain components. BMW's i3 and i8 city cars to be launched in 2013 will have an aluminum chassis housing the powertrain and a CFC passenger cell [Smock 2011]. Similar to Mg, Al components in a multi-material vehicle bring the issues of joining, corrosion, and repair, as well as paints/coatings and recycling.

Advanced high strength steels (AHSS) offer potential weight savings over conventional steel alloys because their exceptional strength and ductility permit efficient structural design [Report 2011]. AHSS could reduce component weight by up to 25%. Because AHSS are generally compatible with current vehicle materials and manufacturing and assembly processes, they will likely be used to achieve near-term weight reduction. ArcelorMittal's "S-in motion" project demonstrates what can be done with steel to achieve weight savings. The project evaluated its state-of-the-art production process with the "lightweight" steel metals, including dual phase, complex phase, Trip 780, Martensitic, press-hardened steel, and other AHSS, and provides a portfolio of material/design options for body panels (front, side, rear), chassis, doors, B-pillars, and front and rear rails. Although this two-year project came out of nearly 10 years of work, ArcelorMittal sees that the creation of new alloys and other developments can provide even more weight savings [Buchholz 2011].

Lotus Engineering has a mission to achieve performance through lightweight materials, but not through intensive use of any specific lightweight material [Gehm 2010]. Probably most known as a manufacturer of premium niche and racing cars, Lotus Engineering also provides consultancy to the automotive industry. Using the 2009 Toyota Venza crossover utility vehicle, Lotus Engineering conducted a study on replacing the mild steel with high strength steels (HSS) in the BIW, and realized a 16% weight savings and 2% cost savings [Gehm 2010]. The HSS BIW contained 89% HSS by weight, which it believes to be feasibly production-ready for 2017, considering that the 2010 Mercedes-Benz E-Class reportedly uses 72% HSS [Gehm 2010].

In the steel industry, current research is targeted to developing the third generation of AHSS that will have properties intermediate to the first and second generations and lower cost (Fig. 8.3) [Report 2011].

In a life-cycle assessment (LCA) of different lightweighting materials (aluminum, glass fiber reinforced plastic, and CFC) for exterior body closure panels, Overly et al. [2002] found that CFC was the least environmentally burdensome material in 9 of 14 categories evaluated. The LCA considered potential environmental impacts from the production, use, and disposal of the closure panels. However, recognizing that continuing advancements will be made in all automotive materials, the authors recommend that: 1) increasing the recycled content of automotive wrought aluminum be evaluated; 2) ultralight steel be evaluated; and 3) part repairability and replaceability be considered.

Aside from the need for CFCs to continuously improve to stay competitive with alternative materials that are also improving, there is the need for CFCs to continue

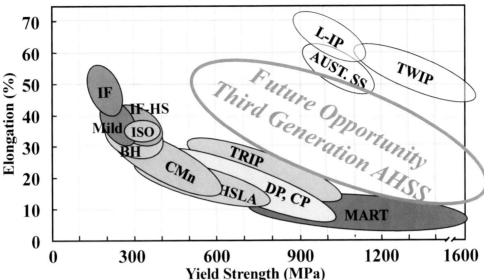

New 3rd Generation AHSS Grades Offer Another 10 % More Mass Reduction with AHSS, that is over 35% for 2017-2025 time period.

Figure 8.3. AHSS properties.
(Courtesy of U.S. Department of Energy)

to strive for compatibility with these alternative materials (as they too advance) to which they will inevitably be joined in the vehicle. Joint design, welding/bonding materials, joint strength, material compatibility (thermal expansion, galvanic corrosion), and protective coatings will need to continuously evolve to enable the integration of multi-materials in a vehicle design.

Partnerships between the advanced materials industries would likely lead to game-changing developments for the vehicle industry and its inevitable advanced multi-material vehicle. The DOE funding for two multi-year programs to demonstrate manufacturability of a lightweight, multi-material vehicle could spark such partnerships [EERE 2011]. It will be very interesting to see the outcome of those programs.

8.3 Industry and Public Acceptance

The automotive industry appears to be on the cusp of revolutionizing vehicle design to finally incorporate advanced materials such as CFCs in mainstream vehicles. A major impetus of this movement is the federally regulated CAFE standards mandating significant improvement in fuel economy across vehicle fleets. If the "stick" is CAFE fines, the "carrot" is a public preference for lighter weight vehicles that will cost less to drive and provide the same or better performance, in durability, maintenance, and repair, and even in crash.

Both the industry and public demand a lot from the materials and components used in vehicles. Industry expects the materials to be inexpensive, readily available, and fast to process, and it expects the components to be easy and fast to assemble, provide good aesthetics, to be durable in terms of mechanical, chemical, and environmental loads, and in crash, to transfer load, dissipate energy, and/or fracture in a non-harmful manner. The public expects the visible parts of the vehicle to look and feel good and withstand the rigors of everyday use, and expects all parts of the vehicle to be robust and durable, easy and not expensive to maintain or repair, and to provide protection in accident, whether a low-speed fender-bender or a rollover crash.

As any new material becomes integrated into mainstream vehicles, there is a possibility for derailment. Some possible derailment events could be unseen by the customer, such as an assembly procedure that is too costly or time-consuming for an automaker to maintain, or an in-field issue with a component or assembly performance that requires costly repair or replacement. Such events would not necessarily derail others in the industry, unless the issue was common across the industry. However, a detrimental event that is publicized can affect public perception, which in turn can cause the demise of a new material or technology.

To expound, consider the ethereal expectations we have of our vehicles. Not only do they need to protect us in a primary crash event, they must retain enough integrity to protect us in secondary crash events. These secondary events, although foreseeable, are not predictable. An example is a T-bone impact into the side of a vehicle, which causes a subsequent secondary rollover event. The levels of protection that can be reasonably provided in primary or secondary events, and for what magnitude of event, are heavily studied and evaluated by the industry. But, what the industry can provide and what the public thinks is being provided may not be the same. This disconnect exists to some extent in current vehicle design with more conventional materials, but could be much more prominent with the use of a new material — and in the case of CFCs, one that performs differently than metal. The general public is not well informed about automotive carbon fiber composites and their characteristics, and perceptions can range from thinking it is an extremely high-performance aerospace material to thinking it is just like a plastic shower insert or fiberglass canoe. Public perception is powerful; consider the demise of the sound and successful passenger-carrying airship technology because of an unfortunate series of events that led to the horrific Hindenburg crash.

While on the path to introducing a new material such as CFCs, the automotive industry has an opportunity to educate the public about the material and what it can or cannot do. It has the opportunity to convey that although CFCs are complex material systems, there are many aspects that are understood, controlled, and checked, from the processing to modeling to testing and quality control, done using standardized or industry-accepted practices.

In our litigious society, an unwarranted accusation of inadequate performance or defective part can unfairly sway public perception against a new material or product. An industry that is diligent about working together to develop accepted practices and standards to ensure quality and predictability in our CFC components and assemblies can create a vehicle design revolution while bringing the public along for an educational ride. The goal is to take innovation to implementation, and eventually to longevity.

8.4 Closing Thoughts

This is an exciting time for the carbon fiber composites and automotive industries. The current need to drastically lightweight the U.S. vehicle fleet in the next few years provides a great opportunity for CFCs to find prominence in mainstream vehicles. Their advantageous high specific modulus and strength can result in weight savings up to 60% compared to conventional steel designs. However, before CFCs find prominence, significant inroads in reducing the relatively high material and fabrication costs, long part-to-part cycle times, and slow assembly/attachment to other vehicle components will need to be made. Progress will also be needed in the areas of damage detection, repairability/replaceability, and recycling.

The purpose of this book was to highlight current activities surrounding automotive carbon fiber composites and the anticipated direction of developments in the next 5–10 years. The combined chapters provide a high-level report as opposed to technical treatises, and should prepare the reader for meaningful discussions with composites engineers and technicians, fiber suppliers, resin suppliers, tool and equipment manufacturers, as well as business development and lifecycle workers. The possibilities of carbon fiber composites in automotive applications are plentiful — and more promising than ever before in the history of the automobile.

References

Automotive World Ltd. April 4, 2011. "OEM Tracker: BMW Group." *Automotive-World.com.*

Buchholz, Kami. 2011. "Materials 'S-in motion' project." *Automotive Engineering International Online.* January 20, 2011. (http://www.sae.org/mags/AEI/9299 Accessed June 8, 2011). SAE International, Warrendale, PA. 2011.

Canis, Bill and Brent D. Yacobucci. 2010. "The U.S. Motor Vehicle Industry: Confronting a New Dynamic in the Global Economy." *Congression Research Service* no. 7-5700 R41154 (http://www.fas.org/sgp/crs/misc/R41154.pdf Accessed August 18, 2011).

"EERE News Department of Energy Awards More Than $175 Million for Advanced Vehicle Research and Development." 2011. no. August 10, 2011. (http://apps1.eere.energy.gov/news/progress_alerts.cfm/pa_id=590 Accessed August 10, 2011).

Gehm, Ryan. 2010. "Even Lotus considers high-strength steel a lightweight option." *Automotive Engineering International Online* (July 6, 2010). http://www.sae.org/mags/aei/mater/8512 SAE International, Warrendale, PA. 2010.

———. 2011. "Finding the right balance." *Automotive Engineering International Online* no. April 5, 2011:43-45. SAE International, Warrendale, PA. 2011.

Klier, Thomas and Daniel P. McMillen. 2008. "Evolving Agglomeration in the U.S. Auto Supplier Industry." *Journal of Regional Science, Vol. 48* no. 1:245-267. 2008.

NREL, National Renewable Energy Laboratory. 2010. "Materials Technologies: Goals, Strategies and Top Accomplishments." no. August 2010 (http://www1.eere.energy.gov/vehiclesandfuels/pdfs/materials_tech_goals.pdf Accessed October 7, 2011).

Overly, Jonathan G., Rajive Dhingra, Gary A. Davis, and Sujit Das. 2002. "Environmental Evaluation of Lightweight Exterior Body Panels in New Generation Vehicles." SAE Paper No. 2002-01-1965. SAE International, Warrendale, PA. 2002.

Platzer, Michaela D. and Glennon J. Harrison. 2009. "The U.S. Automotive Industry: National and State Trends in Manufacturing Employment." *Federal Publications.* no. Paper 666 (http://digitalcommons.ilr.cornell.edu/key_workplace/666).

"Report 10 Lightweighting Material." 2011. U.S. *Department of Energy, Energy Efficiency & Renewable Energy* no. DOE/EE-0577 January 2011.

Sloan, Jeff. 2007. "From the Editor - 10/1/2007." *Composites Technology.* October 2007 (http://www.compositesworld.com/columns/from-the-editor---1012007 Accessed August 17, 2011).

———. 2011. "A part-per-minute start?" *Composites Technology* April 2011 (http://www.compositesworld.com/columns/a-part-per-minute-start Accessed August 17, 2011).

Smock, Doug. 2011. "New BMW Line Features Aluminum, CFRP." *Design News* April 25, 2011 (http://www.designnews.com/article/517938-New_BMW_Line_Features_Aluminum_CFRP Accessed June 8, 2011).

Sullivan, Rogelio, A. 2006. "Automotive Carbon Fiber: Opportunities and Challenges." *JOM* no. November 2006.

"Survey: Composites designers have a lot to learn." August 2, 2011. *Composites World* (http://www.compositesworld.com/news/survey-composites-designers-have-a-lot-to-learn).

Warren, C. David, Felix L. Paulauskas, Fred S. Baker, C. Cliff Eberle, and Amit Naskar. 2008. "Multi-Task Research Program to Develop Commodity Grade, Lower Cost Carbon Fiber." 40th International SAMPE Technical Conference - Memphis, TN - Sep 8 - 11, 2008.

About the Author

Dr. Rehkopf is the Senior Researcher at Plasan Carbon Composites, the leading Tier 1 producer of carbon fiber composites for the U.S. automotive industry. Working out of Tennessee at their facility at Oak Ridge National Laboratory and out of Wixom, MI in their Customer Development Center, she leads Plasan's research in carbon fiber composite constituents and constructions, focused on commercialization. Dr. Rehkopf is also principal investigator of a three-year DOE sponsored project on predictive modeling of carbon fiber composites in automotive crash applications. She is a Civil Engineer with Bachelor's and PhD degrees from the University of Waterloo in Ontario, Canada. She has spent her career working in the field of mechanics of materials. Prior to joining Plasan she worked in research at Ford Motor Co. and as a consultant at Exponent, a failure analysis consulting firm.